Understanding Cellular Radio

For a complete listing of the *Artech House Mobile Communications Library,*
turn to the back of this book.

Understanding Cellular Radio

William Webb

Artech House
Boston • London

Library of Congress Cataloging-in-Publication Data
Webb, William, Dr.
   Understanding cellular radio / William Webb.
        p.   cm. — (Artech House mobile communications library)
      Includes bibliographical references and index.
      ISBN 0-89006-994-8 (alk. paper)
      1. Cellular radio.   I. Title.   II. Series.
   TK6570.M6W43   1998
   621.3845'6—dc21   98-2921
                                                                      CIP

British Library Cataloguing in Publication Data
Webb, William
   Understanding cellular radio—(Artech House mobile communications library)
   1. Cellular radio
   I. Title
   621.3'8456

   ISBN 0-89006-994-8

Cover illustration by Eli Cedrone

© 1998 ARTECH HOUSE, INC.
685 Canton Street
Norwood, MA 02062

International Standard Book Number: 0-89006-994-8
Library of Congress Catalog Card Number: 98-2921

10 9 8 7 6 5 4 3 2 1

Contents

Preface

Objectives

As I have increasingly moved away from the world of research and into that of management and business, it has become apparent to me that there are many people from a nontechnical background who need to understand cellular systems in more detail in order to be able to perform their job, be it analyzing financial investments, running call centers, or marketing cellular services. The problem is that cellular and wireless local loop systems have become increasingly complex over the years and understanding these systems is becoming difficult even for engineers. Indeed, I doubt whether there is anyone in the world who can really claim to understand fully all the parts of the *Global System for Mobile Communications* (GSM) in detail—some will be experts in speech coders, others in error correctors, security algorithms, or network protocols. The complexity of these systems, coupled with the dramatic growth of cellular and the resultant influx of people into the industry, has resulted in increasingly frustrated individuals seeking to better understand these systems and has spawned a plethora of training courses. Unfortunately, these courses often do little to help. Those performing the training are themselves sometimes ill-informed about cellular radio design and simply pass on inaccuracies to their class.

Giving training courses myself and interacting with the nontechnical staff daily, this state of affairs became clear to me. In answering the questions raised during these exchanges, it became apparent that I could do much to inform these nontechnical individuals about wireless and, particularly, cellular technology. The result of these training courses and discussions is this book. I hope that it helps you understand cellular radio.

Acknowledgments

I would not have been in a position to write this had it not been for Professor Raymond Steele, who taught me most of what I know about the world of cellular radio and ensured, with his annotations of "NCE" (meaning "not clearly explained") across my early texts, that my writing style developed into one suitable for such a book. This book has also become a much more readable one as a result of those nontechnical individuals who acted as reviewers to ensure that I had not left any concepts insufficiently explained. Key amongst these were my wife, Alison, and my brother, Matthew, who ensured that the book was approachable. I suspect that there are not many technical books where close family have played such an interactive role!

Part I

Wireless Communications Systems

1

Introduction

1.1 Why a book simply explaining technical aspects?

There are many excellent books covering cellular radio and telecommunications systems, ranging from introductory texts through to detailed descriptions of esoteric parts of cellular radio networks. However, all these books are written primarily for engineers or for students studying mathematically related disciplines. There seems to be an absence of texts explaining the technical aspects of the system without any mathematics and in a manner that can be understood by those with no previous exposure to engineering concepts. However, there is an increasing number of people who work alongside engineers within cellular networks who need to understand the basics of the system operation but do not come from an engineering discipline. Examples of such people include managers, business planners, customer service staff, billing and network management experts, and investment analysts.

This book has been written for these nonengineers. It assumes that the reader does not have a knowledge of engineering terms such as decibels and megabytes per second and explains such terms carefully before using them. It uses each chapter as a building block on which the material covered in subsequent chapters can be based, taking the reader from a low-level understanding to the point where they are familiar with all the key building blocks in the system.

1.2   Scope of material covered

This book covers cellular radio and related systems such as *Wireless Local Loop* (WLL), satellite, and cordless. Most readers will be interested mainly in the cellular aspects and may see the other systems as an unnecessary addition. However, an understanding of these related systems helps to explain some of the key aspects of the design of cellular and the changes likely in cellular systems in the future.

Inevitably, in a text explaining concepts simply to nonengineers, the depth of the content will not be as great as would be the case in some of the other books available on cellular radio. However, surprisingly little has been lost in the "simplification" process and, with the exception of some of the highly detailed engineering aspects of the system, the reader who makes it to the end should have as good an understanding of cellular as many of the cellular engineers practicing today.

In order to enlighten the discussion, examples have been used wherever possible. In most cases these examples have been taken from the GSM cellular system—one of the most successful cellular systems at the moment.

1.3   Acronyms, abbreviations, and other ways to make it complicated

All technical areas abound with acronyms. They are necessary to shorten the space it takes to write complicated descriptions and to make conversation easier. However, they also have the effect of making the area impenetrable to those not well versed in the technology. Wherever an acronym is introduced, the unabbreviated text will always be provided before the acronym. Thereafter, the acronym will be used consistently. Of course, there is a detailed glossary at the back of the book that provides an explanation of what each of the terms means, should you feel like you are getting lost at any point.

1.4   How to read this book

This book has been divided into four parts in order to structure the information more clearly. The contents of each of the different parts is as follows.

Part One. Wireless communications systems   This part provides an overview of what complete cellular systems look like and provides details of why they are structured in this manner. It finishes with an explanation of the basic design of a complete cellular system. By the end of this part, the reader should understand the key building blocks of the cellular system, their function, and the basic reasons for their design.

Part Two. Getting calls to moving subscribers   This part looks at the way in which the mobile radio system knows where the subscribers are, is able to get calls sent to them, and is able to maintain the communications even when the subscriber moves around. Much of the design of a cellular system is based around coping with subscriber mobility; hence, an understanding of this part helps the reader understand much of the cellular architecture.

Part Three. Transmitting the signal   This part moves into the techniques used to ensure that the radio signal travels successfully from the transmitter to the receiver. In this part reside all the detailed areas of processing the transmitted and received signal that are necessary in modern radio systems.

Part Four. Specific radio systems   This part looks at a few key radio systems currently in use, explaining their design, their differences, and which radio systems are used in which situations. This part covers the cellular technologies, the cordless technologies, WLL systems in detail, and the emerging satellite personal communications systems. It concludes with a look at next-generation cellular systems.

This book is designed to be read from start to finish. As new concepts and techniques are encountered, they are explained carefully, but thereafter, used in the same manner as an engineer would use the terms. If the reader has not read the early chapters, then some of the terms in the later chapters may not be clear. Not everyone wants to read a book from end to end, so to assist those who want to read particular parts, key concepts have been introduced in boxed sections, sitting aside from the main text. These boxes are referenced in the glossary. Hence, readers who only look at particular parts should look in the index or glossary if there are any unfamiliar terms in order to find an explanation.

Because the book has been designed to be read from start to finish, it has deliberately been kept relatively short and approachable. Once you have understood this book, you will be well placed to tackle the myriad of more detailed engineering texts on cellular that are available.

Questions have been provided at the end of each section, and model answers are available at the rear of the book. The questions cover the material of the chapter and are designed to help readers assure themselves of their grasp of the important points in the chapter.

1.5   Further reading

This book is only the starting point in gaining an understanding of the complex and multifaceted world of mobile radio systems. There are hundreds of other books available that describe particular aspects in more detail. To catalogue them all would be a long task that would be soon outdated. Instead, at the end of the book, a list of some of the best reference works with details of their level of complexity and relevance is provided.

2

Overview of a Cellular System

2.1 Introduction

This chapter gives you a first look at a cellular radio telephone system. From now on, these systems will be known simply as cellular. It focuses on a complete cellular system while subsequent chapters of the book examine the contents of each of the boxes in increasing detail, helping you understand why the system is put together in the way that it is. By the end of this chapter you should understand what each of the main components of a cellular system is and why it is called a cellular system.

2.2 Why it is called cellular?

Everyone is familiar with the usage of the term "cellular" in describing mobile radio systems. You probably know that it is called cellular because the network is composed of a number of cells. Mobile radio systems work on the basis of cells for two reasons.

The first reason is that radio signals at the frequencies used for cellular travel only a few kilometers (kms) from the point at which they are transmitted.[1] They travel more or less equal distances in all directions; hence, if one transmitter is viewed in isolation, the area around it where a radio signal can be received is typically approximately circular. If the network designer wants to cover a large area, then he must have a number of transmitters positioned so that when one gets to the edge of the first cell there is a second cell overlapping slightly, providing radio signal. Hence the construction of the network is a series of approximately circular cells. This is shown in Figure 2.1.

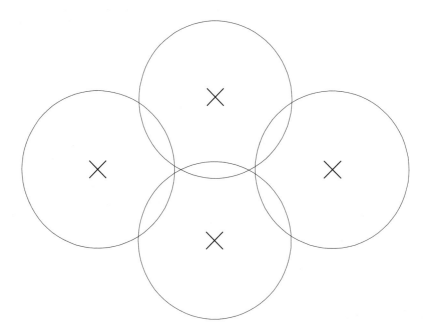

Figure 2.1 Constructing a cellular network as a set of overlapping circular cells.

1. Some radio signals travel much further. So-called short-wave and long-wave signals can travel thousands of kilometers due to their ability to reflect off the atmosphere and back to Earth. At cellular frequencies this phenomena does not take place, so these effects will not be discussed further in this book.

The second reason has to do with the availability of something called radio spectrum. Simply, radio spectrum is what radio signals use to travel through space. Although this is a topic that will be reviewed in more detail in Section 3.2, in essence, whenever a conversation takes place using a mobile radio system, it consumes a certain amount of radio spectrum for the duration of the call. An analogy here is car parks. When you park your car in a car park it takes up a parking space. When you leave the car park, the space becomes free for someone else to use. The number of spaces in the car park is strictly limited and when there are as many cars as there are spaces nobody else can use the car park until someone leaves. Radio spectrum in any particular cell is rather like this. However, there is an important difference. Once you move far enough away from the first cell, the radio signal will have become much weaker and so the same bit of radio spectrum can be reused in another cell without the two interfering with each other. By this means, the same bit of radio spectrum can be reused several times around the country. So splitting the network into a number of small cells increases the number of users who can make telephone calls around the country. This is explained in much greater detail in Section 3.3.

So, in summary, cellular radio systems are often called "cellular" because the network is composed of a number of cells, each with radius of a few kilometers, spread across the country. This is necessary because the radio signal does not travel long distances from the transmitter, but it is also desirable because it allows the radio frequency to be reused, thus increasing the capacity of the network. These are concepts that will be reviewed in later parts of the book.

2.3   System overview

In order to study a complete cellular system, the GSM system is used as an example because it is both one of the most widespread cellular systems in the world today, and one of the best designed. Most other modern cellular systems have chosen to have an architecture similar to it. GSM now stands for Global System for Mobile Communications, but you might think that in that case it should really be GSMC. In fact, GSM was originally an acronym for the name of the committee that was doing the standardization work—called *Groupe Special Mobile*. As GSM was about to be launched, it was decided that such an obscure acronym was not appropriate, so someone managed to find a different meaning for the existing acronym.

The system is composed of three main elements: the switching subsystem, the base station subsystem, and the mobile. In outline, the switching part ensures that when you dial a number you are connected to the right person, the base station part makes sure that the radio communications part takes place correctly, and the mobile allows the user to receive the call. A schematic overview of the complete system is shown in Figure 2.2.

A cellular system cannot exist in isolation. Otherwise, as a cellular subscriber, the only people to whom you would be able to talk would be other subscribers on the same network. The network needs to be connected into the worldwide telephone network so that one can call anyone around the world. This is achieved through the switching system.

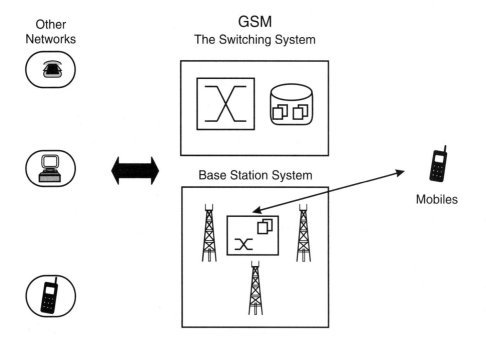

Figure 2.2   Overview of the complete GSM system.

▼   The public switched telephone network

By picking up the phone on your desk, you can make a call to any other phone in the world due to a massive network of wires and telephones spanning the world. This network is referred to as the *public switched telephone network* (PSTN): "public" to differentiate it from a private network (such as the internal telephone system in your office, which only employees can use); and "switched" because whenever you make a call, it passes through a switch, linking you to the person at the other end to whom you want to speak.

The PSTN works by analyzing the number that you dial. Suppose that you are in the United States and you want to speak to a colleague in France. Then you would dial something like 00 33 1 123456. This number is sent from your phone to the local exchange nearest you. The local exchange reads the number from left to right. It only gets as far as 00 when it realizes that this is the code used to request an international call. So it removes the 00, leaving 33 1 123456, and forwards the call to the international switch somewhere in the United States. The international switch starts reading from left to right but only gets as far as 33. The switch has a table of all the international dialing codes, and when it looks in this table it finds that 33 means France. So it removes the 33 and forwards the number 1 123456 to the French international switch. The French switch again reads from left to right, coming across the digit 1. It uses its tables to discover that 1 corresponds to Paris and so forwards the call to the local switch in Paris. Finally, the exchange in Paris looks up the dialing code 123456, discovers that it corresponds to Monsieur Bertillon, and sends the call to him. By this means you can be connected with anyone in the world.

Connecting the cellular network into the PSTN then becomes fairly simple. A cellular network is given its own "local" code (although in this case "local" is a misnomer since the network covers the whole country). If a cellular subscriber dials a number, this passes to the switch in the cellular network that analyzes the first part of the number. If this number is the local code for the cellular network, then the call is destined for another subscriber within the same network; if it is anything else, the switch simply passes it into the nearest local exchange in the PSTN that can deal with it in the normal manner. Similarly, if you try to call someone who is a subscriber to the network from your office phone, then the call will pass to your local switch, which will recognize that the prefix means the call is routed to a particular switch. This particular switch is, of course, the one in the cellular network, but the local exchange knows nothing about whether the switch is for a fixed or mobile network, nor does it need to. (In practice, it needs to know that calls destined for this switch are charged at a different rate so that callers can be charged higher rates for calls to mobiles.)

▲

So a cellular system is composed of three key functional parts dealing with switching the call, making sure the radio signal is transmitted and the mobile radio or receiver.

Now it is time to look into each of those three parts in a little more detail. First, the *base station subsystem* (BSS), which is shown in Figure 2.3, is responsible for making sure that everything associated with the radio

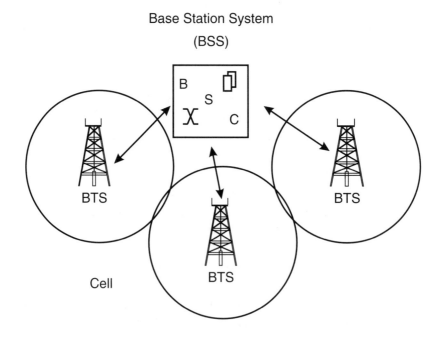

Figure 2.3   The base station subsystem.

signal works correctly. So when the switch sends a message to the BSS saying that there is an incoming call for number 0468 123456, it just wants to hear a message back from the BSS saying "fine, send the voice signal down this particular connection and I will deal with the rest." It does not want to know that 0468 123456 is currently experiencing quite a weak radio signal and it might be necessary to pass him to a different cell.

The BSS mostly consists of a collection of transmitters, known in GSM terminology as *base transceiver stations* (BTSs) where "transceiver" is a shortened version of "transmitter/receiver." Each BTS or transmitter/receiver sits in the center of a cell and radiates radio signals so that mobile users can have conversations. (For simplicity, they will be called transmitters from now on, but it should be remembered that each transmitter has an associated receiver to listen to the signal transmitted by the mobiles.) Transmitters also have a fair amount of computing power, correcting errors in the received signals and encrypting the conversations so that they cannot be overheard. The many functions of the transmitters will be examined in more detail in Part Three of this book. In the case of GSM, there are also *base station controllers* (BSCs) within the BSS that are concentrating points to which a number of transmitters are connected before being connected back along a single line to the switch. This hierarchy simply serves the function of reducing the number of connections into the switch and thus simplifying the role of the switch. The system could function without them by putting more functionality into the switch, however, the design committee decided that this approach would provide the lowest costs. BSCs are a little like middle managers. In a large organization, middle managers mean that instead of 100 employees reporting directly to a senior manager, 10 report to each middle manager and then 10 middle managers report to the senior manager. This sharing of workload tends to make life more manageable.

Next onto the switching subsystem shown in Figure 2.4. This gets a little more tricky, and not all of the parts will be fully explained here. Later chapters of the book will return to some of the concepts discussed here to build on the basic concepts. To understand what is in the switch, first think about what the switch has to do. If you call a cellular subscriber from your desk, the PSTN is going to deliver a request to the mobile switch to deliver a call to 0468 123456. Now think about what the switch is going to do with it.

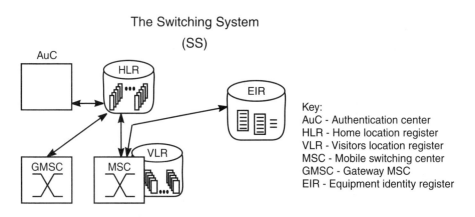

Figure 2.4   The switching subsystem.

All that the switch knows is that, because the number started 0468, it is a subscriber belonging to its network. It has no idea where the subscriber is; it may even have "roamed" to a different network. For example, you can take your GSM phone to a different country and make and receive calls from that country. It does not know if your phone is turned on, whether you have paid your last bill, or whether your phone has been stolen. It needs to do a lot of checking before it can connect you with the cellular subscriber.

All the boxes in Figure 2.4 are there to help the switch, often termed the *mobile switching center* (MSC), with its task. (It is called the mobile switching center, not because the switch can be moved—since it takes up a whole room this is out of the question—but because it switches calls to mobiles.) The switch will send messages to a number of these supporting boxes at the same time and use their responses to decide what to do with the incoming call.

▼   How does a switch switch?

When looking at switches, it is easiest to start with the switches that are in the PSTN, that is, those through which pass all the calls that end up at your home phone. A switch will cover a certain geographical area, with typically something like 50,000 to 100,000 homes in it. From each home comes the copper wires over which your domestic phone service is provided. When you make a local call to someone else within the same switch area (which is the normal definition of a local call), then the function of the switch is to connect your telephone wires with those of the person you are trying to call for the duration of your call and then to release the connection. At the same time the switch may be making connections for any of the other users within its area.

The very first switches were humans manning a switchboard. You phoned the switchboard and told them to which line you wanted to be connected, and they plugged your phone line into the appropriate socket. The first automatic switch was invented by someone called Strowger. Interestingly, Strowger was not an engineer but the director of a funeral parlor. He had suspicions that when people were asking to be put through to his funeral parlor, the switchboard operators were actually putting them through to his competitor, perhaps because the operators had been bribed to do so. To overcome this he invented the first automatic switch, which was known then as the Strowger switch. These old switches worked by laying all the incoming lines across all the outgoing lines in a huge grid pattern. Of course, the incoming lines and the outgoing lines are the same lines depending on whether you originate or receive a call. When a particular connection was required, an electromechanical device known as a relay pulled the lines together at the appropriate crosspoint in the grid structure. This is shown in Figure 2.5 where user E has placed a call to user C and the relay has made a connection in the appropriate point in the grid.

The problem with such switches is that the number of points at which lines cross (and hence the number of relays required) grows very rapidly with the number of incoming lines and hence the switches become expensive. Modern switches are designed on a completely different principle. Simply, all the lines are connected to one ring. This ring can handle many thousands of calls at the same time by using special cables. At each point where a user's wire joins the ring is a device that listens for calls destined for that particular user and when it finds one, directs it off the ring and down the wires. Now the size of the switch can be much smaller. (For those who are interested, in a mechanical switch, for each user there are as many relays as there are other users; hence, the total number of relays equals the number of users times the number of users; for 10,000 users that means 100,000,000 relays. For the new switch the number of

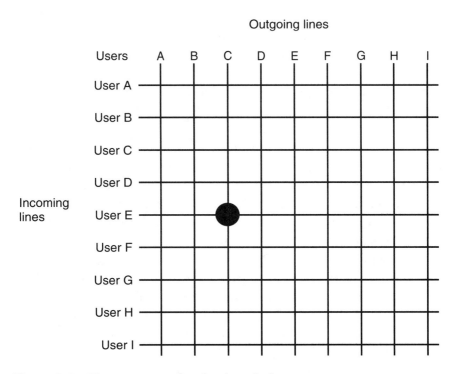

Figure 2.5   The structure of a simple switch.

listening devices is equal to the number of users, just 10,000.) A diagram of a such a switch is shown in Figure 2.6.

Of course, many calls are not local. The switch has a number of connections to other switches, with each of these connections being able to carry many calls simultaneously. If the switch decides that the call is not local, it connects the wire of the user who made the call with one of these interswitch connections.

▲

To find where the mobile actually is, there is a database keeping track of the mobile's location. In fact, there are actually two, the *Home Location Register* (HLR) and the *Visitors Location Register* (VLR). The basic concept is that every mobile should have an entry in a database alongside which is stored details of their last known location. The network gets information on a mobile's location periodically because every now and again the

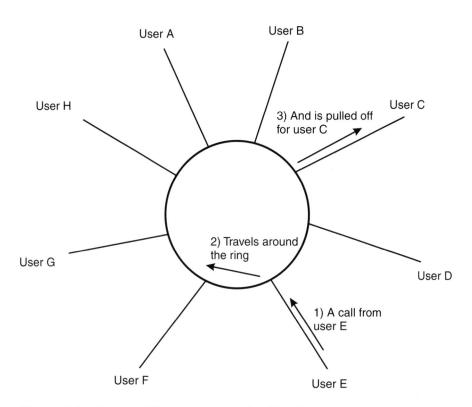

Figure 2.6 The simplified structure of a digital switch.

mobile is instructed by the network to send a message saying "here I am, I'm still switched on." Whenever the network receives this message, it stores details about the cell from which it came in one of the location registers. Hence, the location registers can tell the switch where the mobile actually is (more about this in Part Two). This means that the switch knows to which BSC to send the incoming call.

So why two location registers? Well now consider the case where the mobile phone has roamed to a different country. If there was only one location register, then every time it sent a message to the foreign network, that message would have to be relayed, internationally, back to the home, or subscribed network. An alternative, which requires less international signaling is simply to send a message back to the HLR when the mobile first enters the new country saying that the mobile is in a different network and that any calls for that mobile should be forwarded to the different network until further notice. The database in the different network, known as the VLR, is then responsible for keeping track of where the visitor is while it is in the foreign network. As with the BSCs, the network could work without VLRs, however, the presence of VLRs makes the network slightly easier and more cost effective to operate.

In fact, although this explanation has the benefit of simplicity, it is not quite accurate. In the GSM system, each switch or MSC has its own VLR. When a mobile moves from the area covered by the cells connected to one MSC into the area covered by cells connected to a different MSC, the new MSC loads the information from the HLR into its VLR. This method has all the advantages explained previously and also reduces the processing time for calls being set up even when the mobile is within its home network through storing all the necessary parameters locally.

Another function of the switch is to check whether the mobile is stolen and, if so, prevent any calls being made to or from it. The *Equipment Identity Register* (EIR) is simply a database keeping track of stolen mobiles, in much the same manner that there are databases keeping track of stolen credit cards. Whenever a call is made to or from a mobile, the switch checks with the EIR to make sure that the mobile is not stolen before allowing the call to proceed.

Next, the switch is in charge of authenticating the mobile. This is the process of making sure that the mobile is who it says it is. Otherwise, Paul's mobile could pretend to be Peter's mobile, with the result that Peter would receive the bills relating to Paul's calls. Authentication is described in more detail in the next part. Suffice it to say that the *Authentication Center* (AuC) contains a secret number known only to the mobile. The AuC gives the switch a message to send to the mobile and the response it would expect from the mobile. The mobile takes the message and combines it with the secret number to devise the response. If the mobile sends back the expected response, the switch considers it to be the correct mobile.

The final box in Figure 2.4 is the *gateway mobile switching center* (GMSC). This is effectively an interface between the internal cellular network and the PSTN. The need for a GMSC is explained as follows. Each MSC only has sufficient capacity for a certain number of subscribers, perhaps 500,000, depending on the type of MSC. Each MSC will be responsible for a particular area of the country, for example, there might be one

MSC for New York. In a typical GSM network there might be quite a number of MSCs, perhaps 10 or more. When there is an incoming call from the PSTN, the PSTN has no way of knowing to which MSC it should send the call—the dialed number does not help since the mobile could actually have moved anywhere in the network. So the PSTN needs a single point in the network to which it can send all its incoming calls. This point is called the GMSC. The GMSC is connected directly to the HLR (they are normally in the same building), so it can quickly determine where the mobile is and then forward the call to the MSC that is responsible for the area where the mobile is currently located.

Since it is the point of interconnect with the PSTN, outgoing calls from mobiles destined for the PSTN are also passed through the GMSC. So the GMSC is a special kind of switch, ensuring that calls for mobiles are passed through to the correct part of the network.

Networks also have management systems. These are systems that monitor the functions of each of the components such as the BTS and BSC and warn the operator when they have failed. The management systems also collect statistics such as the number of calls being made to warn the operator when the network is approaching capacity. Finally, they allow the operator to change some parameters of the network, such as the frequency that a particular transmitter is using, without having to visit the transmitter site, making network change simpler and more rapid.

Problem 2.1

Why are cells approximately circular? What do you think are some of the key factors which cause cells to be less than circular in real life?

Problem 2.2

Name the three main components of a cellular network and their key functions.

Problem 2.3

Why is there a need for a VLR?

Problem 2.4

How does a GMSC differ from an MSC?

Problem 2.5

What would happen, if instead of dialing 00 33 1 123456, you accidentally missed the first 0 and dialed 0 33 1 123456?

3

Basic Cellular System Design

3.1 Introduction

It has already been hinted that one of the key issues for cellular systems is being able to accommodate enough users given the limits in radio spectrum. Indeed, this is the single most important issue for almost any wireless system. Radio spectrum is so scarce that systems designers go to almost any length to get more users onto the system. The techniques they adopt to do this form much of Part Three of this book. In this section, the reason why radio spectrum is so scarce is introduced along with a brief explanation of techniques such as auctions to distribute the spectrum. Then the fundamental means by which radio spectrum is shared out amongst the users is described, leading to a simplified "first pass" network design for a cellular system.

3.2   The scarcity of radio spectrum

Radio spectrum is the one fundamental requirement for cellular and wireless communications. By generating an electrical "radio" signal and passing it through an antenna, the signal is radiated in the form of electrical waves that can be received by another antenna some distance away. It so happens that a radio signal is only transmitted when the electrical signal is constantly changing. The simplest form of changing signal is what is known as a sine wave, the shape of which is shown in Figure 3.1.

▼   **Why are sine waves simple?**

It was stated earlier that sine waves are the simplest of all forms of wave and so are used for radio transmission. This statement may not be obvious when square waves, such as those shown in Figure 3.2, might seem simpler to use.

The reason why sine waves are simpler is because sine waves are, in fact, the basic building blocks of all waveforms. A French mathematician called Fourier showed that it is possible to decompose any regular waveform into a constituent set of sine waves. No other waveform can be used in this manner as a basic building block; hence, sine waves are the most fundamental waveforms possible. The mathematics needed to fully understand this is called Fourier Transforms but is beyond the scope of this text.

▲

Most radio transmission is based on generating a sine wave and then changing its parameters in order to send useful information. One of the key decisions is what frequency of sine wave is to be used. That is to say, how many complete cycles or repetitions per second are to be transmitted. Figure 3.1 shows a low-frequency (i.e., only a few complete cycles per second) and a high-frequency (i.e., a lot of cycles per second) sine wave. The frequency of the sine wave has a key bearing on its propagation characteristics, that is, how far it will go from the transmitter, but this is a topic deferred until Chapter 8.

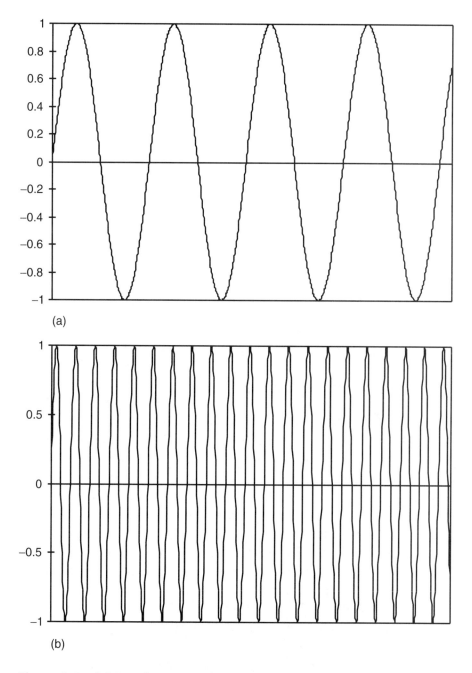

Figure 3.1 (a) Low-frequency sine wave and (b) high-frequency sine wave.

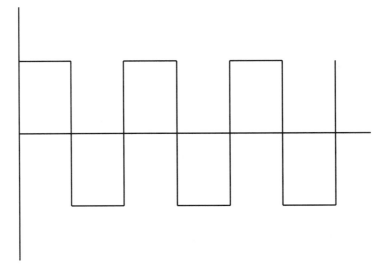

Figure 3.2   A square wave.

Now, if one person uses one particular frequency of sine wave for their transmission and another uses a different frequency, then their transmissions will not interfere with each other. An analogy to this would be if one person talked in a very deep voice and another talked in a very high pitched voice at the same time in the same room. If a filter was then applied that removed all high-frequency sound waves only the person with the low voice would be heard. If a different filter that removed all low-frequency sound waves was applied, then only the person with the high voice would be heard. Bass and treble controls on stereo systems have a similar effect although typically cannot remove an entire voice. However, if both people use the same frequency in the same room, then it is impossible to distinguish between them. They are said to *interfere*.

So everyone who wants to transmit needs a particular frequency at which they can generate their sine wave signal. If two users have the same frequency, then they will interfere with each other and their signals will not get through. However, there is a limited amount of frequency available, and giving each user a separate frequency is not very practical. To find out how much radio spectrum there is, it is necessary to understand the upper limits on transmission of radio signals. But first an explanation of what frequency units such as *megahertz* (MHz) and *gigahertz* (GHz) are.

▼ Hz, MHz, and GHz

You must have come across the term *megahertz* (MHz) when tuning in your radio to a particular FM radio station. This term is composed of two parts, the prefix M and the units Hz. To the units first. Hertz was a German engineer who was one of the first to study radio transmission, and indeed, some still use the term "Hertzian waves" to describe radio waves. In his honor, the term hertz was given to the units of frequency. A sine wave that repeats once per second is said to have a frequency of 1 hertz, written 1 Hz. One that repeats 10 times per second has a frequency of 10 Hz.

Now to the prefix. There is a series of universally agreed prefixes denoting particular numbers. For example, the prefix for a million is "M". Hence 1 MHz is 1 million hertz, which is a sine wave that repeats 1 million times per second. A set of the prefixes likely to be used in cellular is listed in Table 3.1.

The mathematical notation is just a form of shorthand. The term 10^3 means 1 followed by 3 zeros, that is, 1000. The term 10^{-3} means 3 zeros in front of a 1, that is, 0.001 (including the zero before the decimal point). This just saves writing 1,000,000,000 and making a mistake over the number of zeros when 10^9 is shorter and less confusing.

Table 3.1
Mathematical Notation

Prefix	Pronounced	Multiple	Mathematical notation
n	nano	one thousand millionth	10^{-9}
μ	micro	one millionth	10^{-6}
m	milli	one thousandth	10^{-3}
(none)		no multiple	10^0 (=1)
k	kilo	a thousand	10^3
M	mega	a million	10^6
G	giga	a thousand million	10^9

It so happens, as will be shown in Chapter 8, that the highest radio frequency that can be used for mobile radio transmission is around 2 GHz (i.e., two thousand million repetitions per second). Above this frequency, radio waves do not really travel far enough to be of use. It also happens, as will be shown in Chapter 8, that each user actually needs a small band of frequencies all to themselves, around 25 kHz wide, in order to transmit the information that they want to send. If you divide the total amount of spectrum available by the amount each person needs, then you will find that 2 GHz divided by 25 kHz is actually 80,000. But in most developed countries there are millions of people who want to use the radio spectrum to make cellular calls, to listen to the radio, and to watch TV, for example. Clearly there is not enough radio spectrum for each of them.

It was once said that "Spectrum is like real estate—they just don't make it any more." This is quite an apt description. Spectrum is quite like land—there is only a limited supply of it and some parts are more valuable than others. For example, certain parts of the radio spectrum are particularly good for cellular radio and these parts of the spectrum are in much demand from all the companies who would like to be cellular radio operators.

Radio spectrum is usually managed by the Government or their agencies, such as the *Federal Communications Commission* (FCC) in the United States. It is their role to ensure that the rights to use the spectrum are given out fairly to all those who need it and to make sure that no two people are given the same bits of spectrum. Of course, anyone could ignore the government, build themselves a transmitter, and use the radio spectrum. However, they would interfere with the person who had been assigned the right to use the spectrum, that person would complain to the government, and radio detection vans would be sent out to track down and prosecute the person who was transmitting unlawfully.

Giving out radio spectrum fairly is a difficult task. It is a little like trying to fairly distribute the welfare budget; everyone seems to have a valid claim and there is not enough to go round. Recently, some governments have resorted to selling the rights to the spectrum on the basis that the person who is prepared to pay the most must be the person who needs the spectrum most badly. When enough cellular spectrum for around six cellular operators in the United States was auctioned recently, the government received $20 billion in revenue. This shows just how scarce the spectrum is and how much people are prepared to pay for access to it.

What has tended to happen in most countries is that certain parts of the radio spectrum, such as that between 860 MHz and 950 MHz, have been set aside for cellular. This, and some other parts of the spectrum, have then been divided between typically three or four different cellular operators, with the net result that each operator has been given something like 25 MHz of radio spectrum each.

3.3   The use of cells as a means to conserve spectrum

As was discussed previously, a cellular operator has typically been given around 25 MHz of radio spectrum and each individual typically requires 25 kHz in order to make their call. This would appear to mean that there can only be around 1,000 subscribers. However, most cellular operators have millions of subscribers. They achieve this through two concepts. One is frequency reuse and is explained in this section. The other is something called trunking gain and is explained in the next section.

It was mentioned earlier that one of the reasons cellular systems are split into cells is so that the radio spectrum could be reused. Now is the time to discuss this concept in more detail. When a signal is transmitted, it gets weaker the further it moves away from the transmitter in exactly the same manner that a noise gets quieter the further away you move from the source of the noise. Eventually, the signal becomes so weak that it is impossible to detect it at all. If you then go the same distance again away from the first transmitter and another transmitter is installed at this point, it can use the same frequency as the first transmitter because they are so far apart that they do not interfere with each other. At the point between them, the signal from either is just too weak to be detected. This concept of using again the frequency in cells that are far enough apart is known as "frequency reuse" and dramatically increases the capacity of the system, as will soon become clear.

A key parameter in frequency reuse is something called the cluster size or frequency reuse factor. This is best understood with the aid of Figure 3.3. First, a few words about this figure. You will note the cells are

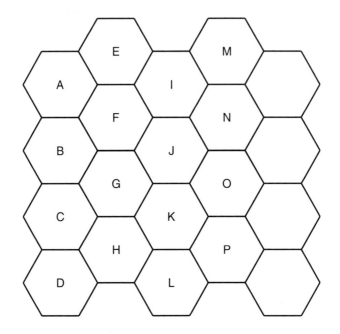

Figure 3.3   A cluster of cells.

shown as hexagons. This is something of a joke in the cellular industry where nobody has seen a hexagonal cell in real life yet. Cells in real life are more or less circular, but unfortunately circular cells do not "tessellate." (A shape is said to tessellate if copies of it can be placed side by side across a piece of paper without leaving any gaps and without any overlap.) A hexagon is quite close in shape to a circle and does tessellate and so most cells are drawn as hexagons to ease understanding.

In Figure 3.3 cells are named A through to P. Now imagine that it is decided to use frequency 1 in cell A. The same frequency certainly cannot be used in cells B, F, or E because they adjoin cell A and so there would always be interference at the edge. Typically, it cannot be used in cells C, G, J, or I either because the interference there will still be too great. However, it could be used again in cells D, H, K, O, N, or M. If we assign our frequencies on this basis, the result might start to look like Figure 3.4.

Now you start to see a pattern emerging. In this particular case it is never necessary to use a frequency higher than seven and the same pattern of a cells with frequency 1 in the center surrounded by frequencies two to

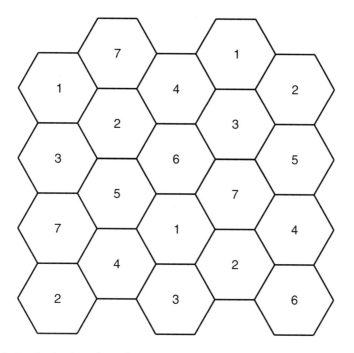

Figure 3.4 A cluster size of seven.

seven can be repeated indefinitely. This becomes clearer if you look at Figure 3.5 where a larger number of cells have been shown and the different clusters are divided by thicker lines. Because only seven frequencies are required, it is said that there is a cluster size of seven.

As explained, a typical system might have 1,000 separate frequencies. So it would be possible to assign frequencies 1, 8, 15, 22,…to the cell that was originally assigned frequency 1. Then frequencies 2, 9, 16,… could be assigned to the cell that was originally assigned frequency 2, and so on. Each cell is assigned one-seventh of the 1,000 channels available. (If the cluster size had been, say nine, then each cell would have been assigned one-ninth of the available channels.) So each cell has around 140 channels, at least for this simple analysis. For a typical medium-sized country, a complete cellular network may have around 2,000 cells and hence a total of $2,000 \times 140 = 280,000$ radio channels. So you can see that from only having 1,000 frequencies it has been possible to generate many more radio channels.

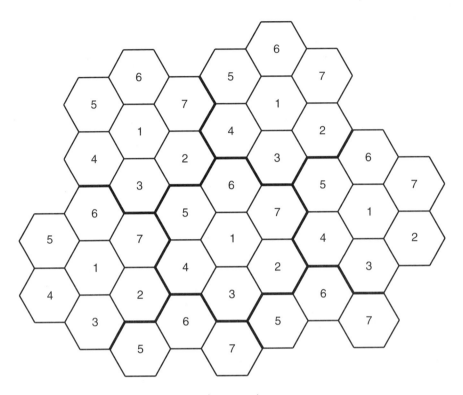

Figure 3.5   The cluster repeated many times.

Even this is not enough. There is still the problem of around 4 million subscribers and only 280,000 radio channels. The next section explains how this problem can be overcome.

▼   Why different systems have a different cluster size

Up until now it has been said that if you travel far enough for the radio signal to become undetectable and then move as far again, then you can put in another cell. It might be imagined that this distance will be the same regardless of the radio system because it is related to propagation laws, not radio system design. In fact, this is a slight simplification. You only need to travel far enough for the radio system to fall to a level where it will not interfere with another radio system. This is the same effect as the two speakers in the same room. You do not need to move the other speaker so far away that you cannot hear them at all in order to be able to have a conversation. Moving them away so that they are much quieter than you, although still audible, is quite sufficient.

Mobile radio systems have a key specification, called the *signal-to-interference ratio* (SIR), that specifies just how quiet the other speaker needs to be before they do not pose a problem. A typical SIR might be around 10. So when you have moved far enough away that the signal level has fallen to a tenth of the minimum signal level that would be experienced at the edge of the cell, if you move as far away again then you can put in another base station reusing the same frequency.

It so happens that the distance you need to move is very sensitive to the SIR. A system with a SIR of 100 would have a reuse distance much greater than one with a SIR of 10, resulting in a much greater cluster size and hence less efficient use of radio frequencies. The actual SIR that a system can tolerate depends on a number of factors, key amongst which is the tolerance of the voice coder to errors on the radio channel and the power of the error correction system that is used (both of these concepts will be explained in Chapter 9). So it can be that different systems can have quite different SIRs hence require quite different distances between the frequencies being reused and hence have quite different cluster sizes.

▲

3.4   Why one channel can serve many users

If you were designing a supermarket, how many checkouts would you have? You might start with the total population of the town. But you know that not all of them will go shopping at the same time. However, you do know that perhaps one in every five will do their shopping on a Friday night. Only a few of those in the supermarket at any time are actually queuing at the checkout. You might go to a neighboring town of similar size and count the number of people going into the supermarket every minute on Friday night. The checkouts need to be able to handle this many people per minute otherwise queues will develop. If you count an average of 10 people going in per minute and time the average person to take 2 minutes at the checkout, then you need 20 minutes of checkout time for every minute of real time, or 20 checkouts.

Then you notice that in one minute 15 people go in while in the next minute only five go in. The average, as you noted earlier, over a period of an hour is still 10, but people do not arrive perfectly evenly spaced apart. What should you do now? You could increase the number of checkout minutes to 30 (i.e., have 30 checkouts) to cope with the peak demand, but then when there were only five people, two-thirds of the checkouts would be idle. It is at this point that you might start thinking that a little science would be useful.

This is an identical problem to the world of mobile radio. Not everyone makes a phone call at the same time; some make a lot, others hardly ever call. If the average user is only on the phone for 10% of the time, then you could share a single channel amongst 10 users. But if you did this you run the risk that two of them will try to use the channel at the same time and one will get a network busy message. If this happens too often your users will migrate to a competitor's network.

This problem was studied in detail by Swedish engineer A. K. Erlang in the early part of the twentieth century. The results he obtained are used in the design of all telecommunication networks. Erlang studied what happened as you varied the number of users that you tried to fit onto a channel and discovered, not unsurprisingly, that the more users you tried to fit onto the channel, the higher the chance that each user would not be able to access the channel, which he termed being *blocked*. He went much further than this. He found that if you had a set of channels, perhaps 10, and you grouped them all together so that if a subscriber wanted to make a call they were able to use any one of these channels that were free, then the probability of them being blocked was reduced. (This is equivalent to being able to use any of the checkouts in the supermarket.)

This seems intuitively reasonable. Say in your supermarket, for no good reason, you decided to split the checkouts into two groups. When a customer came into the supermarket you alternatively assigned them to the left group or the right group of checkouts. Now in some cases, a lot of the shoppers in the right group will finish at the same time and there will be queues in the right group but checkouts free in the left group. If you had not restricted the shoppers in any way, this would not have happened and there would have been fewer queues. In the same way, the more radio channels and users that can be put together in one pool, the less the likelihood that they will be blocked.

In order to formalize his studies, Erlang needed some measure of how many subscribers there were and how busy they were. To do this he invented a measure of telephone traffic subsequently called the Erlang. During a particular period of time (say, an hour) a user generates 1E of traffic if they use the phone continuously. If they only use the phone 10% of the time they generate 0.1E of traffic. If there are 100 users, on average using the phone 10% of the time each, then the total traffic generated is 10E (100 × 10% = 10).

Erlang developed a formula that showed for any number of available radio spectrum channels and any amount of traffic generated, measured in Erlangs, what the probability was that a particular user would be blocked. Actually, this is not typically quite what a wireless operator wants. Most wireless operators decide on what they want the probability of a subscriber being blocked to be. For example, a wireless local loop operator might decide that it would be acceptable for a subscriber to be blocked for 0.1% of all the calls they try and make (i.e., on 1 in 1,000 calls they obtain a "network busy" tone). A mobile operator will typically set a higher blocking probability, perhaps around 2%. This figure is often known as the grade of service. What they then want to know is how the number of radio channels required changes with the amount of traffic generated. A graph showing this for a range of blocking probabilities, P_B, is shown in Figure 3.6.

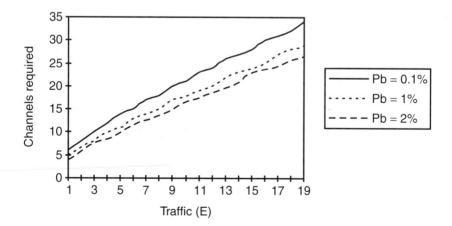

Figure 3.6 Channels versus Erlangs for a range of blocking probabilities.

There are a number of things to note about this. The first, unsurprising, fact is that as the blocking probability gets lower and lower, more and more radio channels are required to handle the same amount of traffic.

The second is the slight steplike nature of these curves that is caused by the fact that only an integer number (i.e., whole number) of channels is possible.

The third is that the number of channels required tends toward the number of Erlangs for high levels of traffic (e.g., using the 2% blocking common in mobile radio systems, for 1E, 5 channels are required—an efficiency of 20%; whereas for 19E, 25 channels are required—an efficiency of 76%). This is the phenomenon discussed previously. In cellular it is known as "trunking gain"—the more traffic that can be trunked together, the more efficient the system. So this implies, with the 140 channels per cell, that almost 140E of traffic can be supported with a low probability of blocking. In fact, the mathematics suggests that something like 120E could be supported with a probability of blocking of less than 1%. Remember that Erlangs are the number of users multiplied by the traffic generated per user.

In the supermarket example, it was noticed that Friday night was particularly busy in the supermarket. If you were designing the supermarket you would design it to take account of this Friday evening peak. The same is true in cellular and wireless systems. Telephone systems have what are called "busy hours." These are the hours during the day when the most traffic is experienced. Actually, most mobile systems are more or less equally busy from about 8 a.m. until 6 p.m., but there are sometimes peaks at around 5 p.m. when everyone is trying to get home and has got stuck in a traffic jam. So, when sizing the network, it is important to understand how many calls the average subscriber makes during the busiest hour. This varies from country to country and depends on the cost of calls (the cheaper they are, the more traffic people generate). Typically, during the busy hour an average user might generate 0.1E, that is, they are on the phone for 10% of the time or six minutes.

With a capacity in each cell of 260E, and an average traffic level of 0.1E per user, 2,600 (260/0.1) users can be supported per cell. With 2,000 cells, a total of 5.2 million subscribers (2,600 × 2,000) can be accommodated. That explains how with only 2,000 radio channels a cellular system can support millions of users.

3.5   Basic network design principles

You now have enough information to attempt some basic cellular system planning. The key objective of planning a system is to build the system to provide enough capacity for the expected number of subscribers but at the lowest possible cost. If you do not provide enough capacity, you will lose subscribers to the competition. If you provide too much capacity, then your network will be more costly than your competitors and they will be able to attract your customers by offering them a lower price. So getting the capacity right is absolutely critical for the success of the network.

Remember from the previous sections that the capacity is affected by the number of radio channels, the number of cells, and the traffic per subscriber. The number of radio channels is fixed depending on what the Government has given you. The traffic per subscriber is more or less fixed—it can be varied by varying the tariff, but you cannot make the tariff too different from your competitors, so there is not much freedom with tariffs. The only variable totally under the control of the network designer is the number of cells to deploy. However, the greater the number of cells, the greater the cost, so the operator will be keen to minimize this number.

There are actually two reasons for putting in cells. The first is the one that has been discussed up to now, that is, to provide enough capacity for the expected number of subscribers. The second is to provide enough coverage of the country, and particularly the populated areas, that the subscribers find having a cellular phone an attractive proposition.

To go any further you need a business plan, or a spreadsheet showing how money will be spent and where the project will come from, and so on. Such spreadsheets often run to hundreds of pages, but a very simplified example is given as follows. The business plan will set the percentage of the country that should be covered and the number of subscribers expected in each part of the country. From the business plan, the following information can be determined and used to determine how many cells are required to provide adequate capacity:

- The number of subscribers;
- The density of the subscribers;

■ The expected traffic per subscriber.

The other important set of parameters is used to determine how many cells are required to provide adequate coverage:

■ The area to be covered;

■ The range of a transmitter;

■ The topography of the area[1].

Having calculated how many cells are required for capacity and how many for coverage, it is necessary to take the higher of the two numbers. Another way of looking at this is that it is necessary to provide a minimum number of cells to cover the target area and, if this does not provide sufficient capacity, then more cells will be needed.

▼   A cellular business plan

The building blocks of the business plan are:

■ The network build costs;

■ The costs of subsidizing mobiles (if applicable);

■ The on-going costs such as maintenance and rental;

■ The predicted revenue;

■ The cost of finance such as interest.

Each of these elements can then be added within the spreadsheet to generate financial information such as expected profit and funding requirements.

Network build costs

The key elements of this are:

1. Base stations;

2. Base station interconnection;

---

1. The topography of the area affects propogation and, hence, coverage.

3. BSCs (if required);

4. BSC interconnection (if required);

5. Switching costs;

6. Operations, maintenance, and billing system costs.

These cost elements are then combined with the number of cells required, as provided by the coverage plan and the network plan, and the rollout planning showing which years the expenditure will be incurred. Table 3.2 shows a typical spreadsheet for a simple system where base station interconnection is not required.

Table 3.2
Example Spreadsheet for Network Cost

Basic assumptions (prices in $000)	
Base station cost	50
Installation	20
Planning	5
Base station interconnect via microwave links	
Cost	70
Installation	10
Switch costs	1000
Switch installation	1000

Year	1998	1999	2000	2001	2002
BSs installed	50	150	100	0	0
Switches installed	1	0	0	0	1
BS total cost ($m)	7.75	23.25	15.5	0	0
Switch total cost ($m)	2	0	0	0	2
Total cost ($m)	9.75	23.25	15.5	0	2

The spreadsheet shows a typical deployment, where most of the base stations are installed in the early years of the network so that coverage is gained as quickly as possible. A single switch is required in the early years, but as the traffic increases in later year an additional switch is required to provide sufficient capacity. With 300 base stations, reasonable coverage of a small country might be expected.

Mobile subsidy costs

The cost of a mobile can be obtained from the manufacturer. There is likely to be a number of different mobiles with different functionality and different costs and each needs to be treated separately. The following information needs to be gathered:

1. The projected number of new units required each year. This will be based on the predicted number of new subscribers expected each year.

2. The projected cost of the subscriber units over the investment period. In most cases, subscriber unit costs can be expected to fall as manufacturing volumes and competition both increase.

3. Whether mobiles are to be subsidized. If they are not, then it is not necessary to take their cost into account.

The on-going costs

Once the network is operational, there is a host of on-going costs that include:

1. Rental for the base station and base station controller sites;

2. Leased line costs where used;

3. Maintenance on all the network equipment;

4. Costs of radio spectrum;

5. Subscriber management costs;

6. General management costs including buildings and facilities;

7. Marketing, sales and subscriber retention costs.

Predicting revenue

At its simplest, the calculation of revenue is just the total subscribers multiplied by the average revenue per subscriber. Unfortunately, both of these figures are subject to a high degree of error.

Subscribers need to be divided into different categories such as business and residential and the total numbers and revenue predicted for each. Tariffing policies need to be consistent with competition and target subscribers. Subscriber numbers need to show the change in subscribers over time and the change in call usage over time. Other factors to include in the calculation are the interconnection costs, taxes, and bad debts. Interconnection costs relate to payments to the PSTN operator when a call originates on the mobile network but terminates in the PSTN operator's network. These vary dramatically from country to country, typically depending on regulatory intervention.

Financing arrangements

This is not intended to be a detailed text on the ways of borrowing money, financial texts are available on that topic. In outline, the options available are:

- Self-funded from reserves within the group;
- Direct funding from banks or other financial institutions;
- Vendor financing from the equipment manufacturer;
- Shareholder funding.

Summary of financial statistics

Having calculated the total expected capital and on-going costs and the revenue, the next step is to compare the two in order to determine whether the selected balance of coverage, service, and tariffs meets the business requirements. The first step is to compare the total expenditure with the total income. This is shown for the simple example in Table 3.3.

The spreadsheet shows an unusually profitable network (deliberately made so to keep the example simple), in practice much longer break-even times are likely. Nevertheless, it adequately and simply demonstrates some of the key financial analysis that is required.

Table 3.3

Example Spreadsheet Showing Summary Financial Statistics

Basic assumptions	
Interest payable on loans	10%

Year	1998	1999	2000	2001	2002
Capital spend ($m)	9.75	23.25	15.5	0	2
On-going spend ($m)	1.8	6.8	11.9	16.2	20.2
Revenue ($m)	4.0	15.5	37.2	70.9	100.7
Profit ($m)	−7.6	−14.5	9.8	54.7	78.4
Borrowing ($m)	−7.6	−22.9	−15.4	37.7	116.2
Interest payable ($m)	−0.76	−2.29	−1.54	0.00	0.00
Bank balance ($m)	−8.4	−25.2	−16.9	37.7	116.2

▲

The calculation of how many cells are required needs to be performed for each part of the country where conditions are different. That is, there is little point in calculating the required transmitter density for a whole state or county in one calculation if that state actually consists of one dense city and otherwise rural areas. The transmitter density would then be too low to provide adequate capacity in the city and too high in the surrounding areas. The calculation should be performed for any areas where the density of subscribers and traffic per subscriber differ significantly.

This might result, for example, in the calculation being performed for the financial district of the city, which has high telecommunications demand, separately from the neighboring area of the city, which might contain businesses with lower demand. The need to perform calculations numerous times is not problematic in itself because the calculation is relatively simple and could quickly be computed on a spreadsheet. The difficulty is in obtaining the input information.

So for each area, the first step is to calculate the number of cells required to cover the area. This is simple; it is just the size of the area, measured in square kilometers (km^2), divided by the area covered by each cell.

The number of cells required for capacity is simply found as the total traffic expected in the area, measured in Erlangs, divided by the capacity of one cell, as calculated previously. So if we expected 26,000 subscribers in a particular city, each generating 0.1E, then there would be a total traffic of 2,600E. If each cell could provide up to 260E, then 10 cells would be needed to provide sufficient capacity. (In practice, the traffic is unlikely to be perfectly evenly distributed between all the cells and a margin of around 20% might be required to account for this effect.)

If more cells are needed for coverage than capacity, then each cell is operating below full capacity—not a particular problem. If more cells are needed for capacity than coverage, then the cell radius must be reduced so that all these cells can fit within the city without overlapping too much with each other. (If they overlap too much, they will interfere more with neighboring cells and the cluster size will have to rise as a result, thus reducing the capacity of each cell.) The cell radius is reduced by reducing the transmitter power, which is analogous to reducing the volume on a stereo when you do not want the sound to go so far.

So that is how you perform basic network design. In coming chapters you will see that there are many more parameters that also need to be selected, but none so important as selecting the correct number of cells.

3.6 Connecting up the network

Before this part of the book is completed, you need a little more information about how the various parts of the network are connected. Clearly, the transmitters (i.e., base stations) need to be linked to the switch somehow, otherwise how else would the phone call get from the switch to the transmitter in order to be sent to the mobile terminal. There are broadly three ways to make this connection:

- Using a communications link leased from the national telephone company;

- Using a microwave point-to-point link;

- Using a satellite link.

Important factors in making the decision include:

- The relative cost, which is affected by the distance of the link, the capacity required from the link, the presence of available infrastructure, and the availability of radio spectrum;

- The desire to avoid using a link provided by the national telephone company, who is probably a competitor;

- The desire to use the same interconnection method throughout the network to reduce costs and complexity.

The concept of the capacity required from the link requires a little further explanation. Clearly, if you want to send two telephone calls down the same link you will require a "higher capacity" link than if you just want to send one phone call. Many years ago your link would need to consist of two pairs of wires rather than one, but nowadays there are many techniques for sending more than one call down the same link. However, the more telephone calls, the more spectrum the link will need if it is a radio link, or the more of the capacity on a copper or fiber optic cable it will use up if it is a cabled link. The capacity required from a link depends on how many telephone calls the base station can support. If it has the capacity for a lot of telephone calls, then a high-capacity link will be required. This might mean using two, rather than one, copper or fiber optic cables. A *post and telephone* (PTO)[2] will charge more for a high-capacity link than a low-capacity link. You will see in later chapters how the capacity of the link can vary and what constitutes high- and low-capacity systems.

In order to understand these factors better, each of the interconnection systems is now explained in more detail.

3.6.1   Leased link

A leased link refers to the leasing of a line from a national or regional telephone operator. The line may not actually be cable, the telephone operators may themselves be using satellite or microwave links, but this will not be apparent to the wireless operator who pays an annual fee for

---

2. The PTO is the national carrier in a country, (e.g., AT&T in the United States).

use of the link that is typically related to the distance over which the link runs and the capacity required from the link. There may also be an initial connection fee to link the transmitter to the nearest access point in the leased link network. Telephone companies typically publish standard prices for such services.

Even when the leased link is the cheapest option, operators may wish to avoid it for a number of reasons. Most notably, links may not be sufficiently reliable or have a sufficiently low error rate; and the operator, for competitive or emotional reasons, may wish to avoid leasing key resources from a competitor whom they might suspect of deliberately providig an inferior service to minimize the competition.

Table 3.4 shows some leased costs as charged by the BT in the United Kingdom at the beginning of 1997.

3.6.2 Microwave links

Microwave links are point-to-point radio devices. A radio transmitter and directional antenna is placed at the transmitter site, pointing to another directional antenna and receiver at the switch site. Directional antennas are devices like the reflector in your car headlamp that cause the

Table 3.4
Example Fixed Link Costs

Link capacity		2 Mbits/s	34 Mbits/s
Connection costs	First link	$18,800	$94,000
	Additional links	$5,900	$47,000
Annual rental	Both ends in London	$7,700	$77,000
	Else		
	If link 15 < km	$11,000 + $600/km	$110,600 + $6,100/km
	If link 15 > km	$16,000 + $300/ further km	$202,000 + $3,000/ further km

signal to be radiated as a narrow beam rather than in all directions. Directional antennas are almost exactly like this; they have a curved metal dish behind the transmitter, reflecting any signal forward. Satellite dishes are a typical example of such a directional antenna. Directional antennas are often used for cellular systems, where it is required to divide the cell into a number of sectors, as will be explained later, and highly directional antennas are used for point-to-point links where there is only a single receiver and no need to radiate signals in all directions.

Microwave links can have a range of up to 50 km depending on whether there are hills in the way and what particular radio frequency is used—the higher the frequency, the shorter the range. The range is much greater than cellular systems because all the power is focused in a narrow beam. This is analogous to a torch projecting a beam further than the light from a bulb of the same power that is suspended from the ceiling.

The cost of microwave links lies in the cost of the link equipment, site rental, and on-going maintenance. Table 3.5 shows typical values 1997 for some of these costs.

The reason for the variation in equipment cost is mostly one of economies of scale, more links are deployed at 38 GHz than other frequencies and so the equipment costs are lower. The site rental varies because it is typically related to the size of the antenna, which affects the "wind loading" on the mast and hence the strength of the structure required to support the antenna. Antenna sizes decrease with increasing frequency. License fee costs vary from country to country but may be high in the congested bands to encourage users to move to higher bands. Due to

Table 3.5
Typical Costs in 1997 for Fixed Links in Various Frequency Bands

Frequency	Equipment cost ($)	Annual site rental ($)	License fee ($)	Annual maintenance ($)	Total cost ($)
4 GHz	80,000	4,800	1,600	2,000	164,000
13 GHz	80,000	4,800	1,600	2,000	164,000
22 GHz	60,000	3,000	1,000	1,500	115,000
38 GHz	48,000	2,400	800	1,200	92,000

historical reasons, the lower bands are currently the most congested. Maintenance is typically calculated as a percentage of the capital cost, and hence the higher cost links will also have a higher maintenance cost.

The cost of a fixed link is broadly not related to either the distance of the link or the capacity required. Hence, when comparing fixed link costs with leased line costs, the result is typically that shown in Figures 3.7 and 3.8 for low-capacity and high-capacity links, respectively. The fixed link costs increases in steps as the range increases because progressively lower frequency systems are required to provide sufficient range, which, as Table 3.5 showed, are more expensive. For

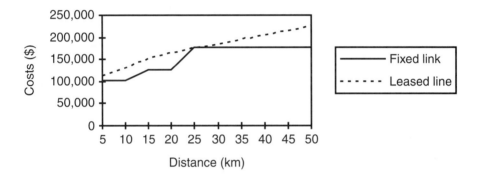

Figure 3.7 Fixed link versus leased line costs for a low capacity link.

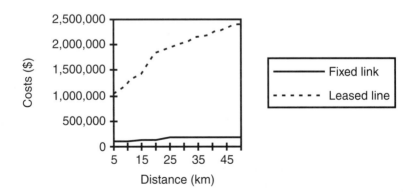

Figure 3.8 Fixed link versus leased line costs for a high-capacity link.

this example, fixed links are always less expensive than leased lines. However, a change in leased line tariffs could instantly change this comparison.

3.6.3 Satellite links

Typically, satellite links are more expensive than fixed links and delay the signal much more than the other types of link because the distance to the satellite is much greater than the distance between transmitter and switch, and hence the signal takes longer to travel this distance. This delay is perceived as echo by the user and is highly undesirable. However, in situations where the cell site is so remote from the switch that a number of hops on a fixed link would be required and there are no available leased lines, satellite links may form the only viable option.

To establish a satellite link, capacity must be leased from a satellite operator, such as Inmarsat, who will provide assistance in installing the dishes and appropriate transmitter and receiver equipment. The cost of a 2-Mbits/s leased line over satellite is currently around $480,000 per year, depending on location, so a 10-year through life cost would be around $4.8M, which is substantially more expensive than the other options.

Problem 3.1

If you are on the phone for 12 minutes of the busy hour, how many Erlangs do you generate?

Problem 3.2

What is 1,000,000 Hz written in a similar notation to 1 kHz and also to 10^3 Hz?

Problem 3.3

What are the three key factors that affect the system capacity, and which are within the control of the operator?

Problem 3.4

What is trunking gain, and what are its implications to the cellular operator?

Problem 3.5

Is a large or small cluster size best, and why?

Part II

Getting Calls to Moving Subscribers

THE KEY DIFFERENCE between the fixed telephone networks that deliver calls to your house and cellular networks is that in cellular networks the user can take the call while on the move. Some of the implications of mobility have already been seen with the study of the HLR and VLR in Chapter 2. In this part, a much more extensive look at all the implications of mobility is provided. The discussion in this part is heavily based on the GSM system, probably the system with the most advanced and well-planned mobility features to date. Current thinking is that the next generation cellular systems will adopt a near-identical means of coping with mobility to GSM.

This part is structured as follows. Chapter 4 looks at how the network keeps track of where the users are, a prerequisite to making calls to the users. Chapter 5 looks at how, once a user has been located, a call can actually be placed through to the user. Chapter 6 examines what happens when a subscriber moves from one cell to another during a call; while Chapter 7 looks at international roaming, security, authentication, and a few other topics relating to mobility.

4

Keeping Track of Where the Users Are

4.1 Introduction

In Chapter 2, some of the concepts of making calls to mobile subscribers were briefly introduced, including the need to keep track of where the users are and being able to hand them over to different cells as they move around during the call. This section looks into the task of keeping track of the mobiles in somewhat more detail.

4.2 The concept of location areas

You may remember that, when switched on, a mobile phone has to periodically inform the network of where it is. In fact, the demands on the mobile are slightly greater than this. It must also tell the network whenever it moves to a different area so that the network always knows exactly where it is. But what is meant by a different area, and how does the mobile know when it is in a different area?

The smallest area that can be distinguished by a cellular system (and by the mobile) is a single cell. In GSM, each cell broadcasts a unique identity that all the mobiles in the cell can receive. So the mobile knows in which cell it is, but unless it transmits something back, the transmitter does not know which mobiles are in which cell. It would be easy to have a system whereby whenever the mobile moved to a different cell, which it detected because it found a different cell identification number, it would send a message to the transmitter saying that it was now in this particular cell. However, the problem with this approach is that in some cities, the cells are quite small, perhaps only a few hundred meters across. If a mobile was moving quickly, it would be forever sending messages to the network telling it where it was. This would have two detrimental effects. The first, key effect, to which we will continually return throughout this book, is that these transmissions use up radio spectrum and radio spectrum is so scarce that everything possible is done to minimize its use. The second is that whenever a mobile transmits, it uses up battery power. Continually reporting its location would rapidly drain the battery.

But if the mobile does not tell the network exactly in what cell it is located, how will the network be able to find the mobile? The solution to this problem is to group a number of cells together and call them a *location area*. A location area might group, say, 10 cells together. Now instead of broadcasting the cell identity, the transmitter broadcasts a location area identity that will be the same for a number of adjacent cells. Mobiles only transmit messages to the transmitter telling it that they have changed area (so-called "location area update messages") if they detect a different location identity code on moving to a new cell. In this way, the location update traffic is reduced by approximately the number of cells in each location area (e.g., if there are 10 cells in a location area, then the location update traffic will only be one-tenth of the case where each cell was a separate location area).

The problem now is that the network only knows the location of the mobile to the accuracy of the location area (say 10 cells). So when it has an incoming call for that mobile it has to send out a message in each of the 10 cells, informing the mobile that there is a call for it and then wait to see in which cell it detects a response from the mobile. This sending of a message is called *paging* and is discussed in the next chapter. But, of course, sending a paging message uses up radio spectrum too, and having to send it in, say, 10 cells uses up more spectrum than having to send it in only one.

An optimum size of location area needs to be found where the combination of location update and paging messages is at a minimum. This depends on factors such as the size of the cell, the amount of incoming calls for the users, and the speed at which the users are moving around the cells, and so will differ for different parts of the network. That is why there is no fixed size for a location area; the operator can make them as big or small as they like. They can also change them quite easily once the network is built and so can optimize the location areas once they get some idea of the traffic load. However, simply, if a mobile changes to a different cell more often than it receives an incoming call, then the use of location areas reduces the total use of the radio spectrum, leaving more for calls between subscribers.

4.3   Attach and detach

So the network now knows where the mobile is, or at least where it last was when it reported its location. But what if the mobile got turned off? To some extent the network does not care about this. If the mobile is turned off, then it will not be able to send a call to it anyway so it does not really matter where it is. The problem here is that in order to find that the mobile is not responding it is necessary to send a paging message in the location area where the mobile was last reported, await any response, and then decide that the mobile is no longer available. As mentioned in the previous section, sending a paging message uses up radio spectrum, and everything possible is always done to minimize the use of radio spectrum. One way to prevent the network having to send a paging message is for the mobile to tell the network when it has been switched off. This is what happens in a GSM network.

When you switch off your GSM mobile, it does not actually turn itself off for a few seconds after you switched it off. First, it sends a message to the network saying it has been turned off. This is known as a "detach" message. The network remembers that your mobile is now switched off and if there are any incoming calls, immediately responds to the caller with a message saying that the mobile is not available, saving radio spectrum and giving the caller a faster response. The knowledge of whether a mobile is on or off is kept in the HLR so that when the network looks for a mobile that has an incoming call it can be told that it is turned off.

Of course, the network now needs to know when your mobile is switched back on again. When you turn your mobile on, it sends an *attach* message to the network saying that it is now turned back on again and ready to receive any calls. This information is forwarded to the HLR along with the location area in which the attach message was received.

4.4   Periodic location updating

It would appear that the network now knows everything it needs to know—whether the mobile is on or off, and if it is on, where it is located. Unfortunately, everything does not always work as well as this. The user might take the mobile into an area where there is no coverage. By the time that the mobile has worked out that there is no coverage it is unable to send a message to the network to this effect, because, by definition, if there is no coverage it is not possible to send or receive a message from the network.

In this case, the network still thinks that the mobile can be contacted in the location area it last reported. Another circumstance that can lead to this situation is the mobile's battery running out. Finally, and a little more disastrously, the computer containing the location information, the HLR and the VLR, might crash and lose all their information. For all these reasons, mobiles that are turned on are instructed to periodically send a message saying where they are, even if they have not moved into a different location area. This is known as "periodic location updating."

Yet again there is another trade-off here. If periodic location updates are required very frequently, then radio spectrum will be wasted for lots of location updates, most of which contain no new information. If the location updates are very infrequent, then they will take a long time to detect that the mobile has left the coverage area and a paging message might be sent to it in the meantime. The operator has to select an appropriate time, typically something like an hour, that is broadcast to the mobile from the transmitters. If the network has not heard from the mobile after the elapsed time indicated, it marks it as turned off and awaits a signal from the mobile before incoming calls will be routed to the mobile.

The network now knows everything it needs to. It knows where the mobile is and it knows whether it is turned on or turned off. In the next chapter, the means of getting calls to mobiles after having located them is described in more detail.

4.5   The mobile in a call

When the mobile is in a call, the network has excellent information about it. It knows exactly in what cell it is because it is transmitting and receiving from the mobile in that cell. It has a channel to the mobile, so it can ask it for any information that it might require.

When the mobile is in a call, this is noted in the location registers. If another call comes in for the mobile during this time, the note in the location register quickly informs the network of this fact. In GSM there is a facility known as "call waiting" where users engaged on a call can be told that they have another incoming call. If the user has subscribed to this facility, then the network directs the call to the correct cell and sends a message to the user on the channel that they are currently using that they have another phone call. They can then chose whether they wish to accept it.

Problem 4.1

What would be the effect of increasing the size of a location area from 10 cells to 100 cells?

Problem 4.2

If a large cell is replaced by a number of smaller cells (this is a technique often used to enhance capacity) what would the network planner do to the location areas?

Problem 4.3

Why are periodic location updates necessary?

Problem 4.4

Why does the mobile not turn itself off immediately when you press the OFF button?

Problem 4.5

Where does the network store information as to whether the mobile is turned on or off?

5

How Calls are Made Between Two Users

5.1 Introduction

The network knows where a mobile is, at least to within a group of cells. In this chapter, the procedure it follows to actually make contact with the mobile is described. This is a complex procedure that is a compromise between saving radio spectrum and getting the call established in the shortest possible time. You may have noticed that when calling a mobile there is a significantly longer delay before you hear a dial tone than when calling a fixed line. This chapter will show why this is the case.

5.2 Paging

The concept of paging was already introduced in the previous chapter where it was said that the network sends out a message in all the cells in the location area where the mobile is known to be and awaits a response. Here the paging message is described in a little more detail.

Each cell has a particular channel set aside for the transmission of paging messages. Channel structure will not be addressed until Chapter 11; for the moment it is easiest to imagine that one particular frequency is set aside for the purpose of paging, although as will be seen later this is something of a simplification in GSM. When a mobile is not in a call but is turned on, it is said to be in *idle mode*. (When it is not turned on, it is simply said to be *off*.) When in idle mode, the mobile is far from idle. It does a number of things, one of which is to listen to the paging channel in its cell. Each paging message is of the form "Mobile number 0468 123456, there is a call for you." So the mobile listens to each paging message and compares the telephone number in the message with its own. If they are the same, it responds to the paging message in a manner that is described in the next section.

As you might have come to expect by now, there is a problem. Listening to radio messages drains the mobile's battery. Although the drain is much less than transmitting, nevertheless it is a drain. With manufacturers seeking to produce increasingly small mobiles with increasingly long battery life, this is quite a serious problem. The solution to this was to divide the paging messages into 10 groups, depending on the last digit of the mobile telephone number. Each group is sent in order, starting with the first group up to the tenth group, back to the first group and so on. So all the calls for our mobile with a number whose last digit is 6, would be in the sixth paging group. The mobile then only needs to listen for 10% of the time because it knows that paging messages in the other groups cannot be for it. Each group is known as a *paging subgroup*, which significantly reduce battery drain. However, they also impose a delay on the speed with which the call can be established. The network has to wait until it is time to transmit that particular paging group before it can send the paging message. In the worst case, if the paging message arrived at the cell just after the sixth paging subgroup had been transmitted, the cell would need to wait for all the other nine paging subgroups to be transmitted before it could send the paging message for our mobile. This was decided to be a lesser evil than the problem of battery drain, so when it takes quite a long time to get a ringing tone after you call a mobile, this is one of the reasons why.

5.3   Responding to a paging message

Once a mobile has received a paging message, it needs to get in contact with the network to let it know exactly in which cell it is and that it is ready to receive the call. There is a channel (assume that this is one particular frequency for now) that is set aside to allow mobiles to get in touch with the network. This channel is known as the *random access channel*. The name is quite descriptive, as will now be explained.

The random access channel has two purposes. It is used by mobiles when the user of that mobile makes a call. It is also used by mobiles when they receive a paging message to let the network know where they are.

First, the process of making a call is explained. When the user makes a call, the network has no idea when he might press the send button and so no idea what traffic will appear on the random access channel. It cannot reserve space for any particular mobile because it does not know it needs to make space until the mobile has contacted the network. To the network, messages appear on the random access channel at random.

Randomness always causes problems for radio systems, where normally everything is carefully planned and under the control of the network. The biggest problem on the random access channel is that two users in the same cell might press the send button at the same time. Both mobiles will send a message, called an *access message*, on the same channel at the same time. An access message says to the network "I would like my own radio channel please, so I can set up a call." When both mobiles send a message at the same time these messages are said to *collide*. Earlier it was noted that two users cannot use the same frequency in the same place because they will interfere with each other. But that is exactly what might happen on the random access channel, and the network can do absolutely nothing about it. Instead, it does everything it can to reduce the probability and the detrimental effect of such collisions.

The first thing that it does is ensure that the messages sent by the mobile are very short indeed. The shorter the message, the less the chance that while Paul's mobile is sending its message, Peter's mobile starts to send a message, destroying the last part of Paul's message. This is shown in Figure 5.1. All the mobile sends is eight bits of information that are selected at random by the mobile. What happens to these eight bits will be described in just a moment.

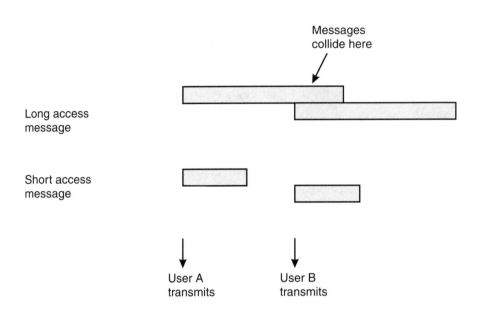

Figure 5.1 The advantages of short access messages.

▼ Binary arithmetic

Mobile radio systems, like computers, use binary as their means of storing and transferring information. An understanding of binary arithmetic will be important for the chapters that follow.

When you count, you use arithmetic that is said to be "base 10" because you use 10 numbers (0 through to 9). When you get to 9, you start back at 0 but also start a second column to the left into which you put a 1, making 10. Each time you get to 9 you go back to 0, incrementing the column to the left. If that gets to 9, when it is incremented it goes back to zero and the next column to the left is incremented. Now imagine the human race was born with only eight fingers and thumbs. Then it would probably count from 0 to 7 but instead of 8 it would use 10_8. The use of the subscript "8" indicates that this number is in base 8 and not the base 10 we normally use.

A computer is a bit like someone born with only two fingers. They count from 0 to 1 but instead of 2 they use 10_2. Below are the binary numbers equivalent to the decimal numbers from 0 to 15. You can see that each time the rightmost digit changes. If this goes from 0 to 1, nothing else changes. If it goes from a 1 to a 0 (equivalent to going from 9 to 10 in decimal), then the next column to the left is incremented. If the increment makes this go from a 1 to a 0, then the next column to the left is incremented.

Converting a binary number to decimal is relatively easy (Table 5.1). Consider decimal for a moment. Take the number 428. To work out what this number means you strictly need to take $(4 \times 10 \times 10) + (2 \times 10) + (8 \times 1)$. That is, as you move to the left by a column, you multiply the number in that column by the base (10) for each time you move left. So you multiply the first column to the left (with 2 in it) by 10. The next one by 10×10 and the next one by $10 \times 10 \times 10$. The same is exactly true in binary. You multiply the first column to the left by 2, the second by $2 \times 2 \ (= 4)$, the third by $2 \times 2 \times 2 \ (= 8)$, and so on following the pattern. So the binary number 1011_2 can be converted to decimal as $(1 \times 8) + (0 \times 4) + (1 \times 2) + (1 \times 1) = 8 + 0 + 2 + 1 = 11$.

Table 5.1
Counting to 15 in Binary

Decimal	Binary
0	0000
1	0001
2	0010
3	0011
4	0100
5	0101
6	0110
7	0111
8	1000
9	1001
10	1010
11	1011
12	1100
13	1101
14	1110
15	1111

Binary operations such as multiplication and addition take place in just the same way as for decimal. Consider the addition of 101_2 with 110_2. This is done as

$$
\begin{array}{r}
1 \quad 0 \quad 1 \\
+ \quad 1 \quad 1 \quad 0 \\
\hline
1 \quad 0 \quad 1 \quad 1
\end{array}
$$

Starting at the right, $1 + 0 = 1$. The same is true for the second column from the right. In the third column from the right $1 + 1 = 0$ carry 1 (in the same way that $9 + 1 = 0$ carry 1 in decimal). Hence the result is 1011_2. Binary multiplication is equally simple.

An operation that is used quite a lot in mobile radio is modification by a "mask." In this operation, a sequence of binary digits carrying the user's information is combined with another sequence of binary digits called a mask.

Consider the binary sequence 10111001 modified by the mask 10101010. The operation looks like

$$
\begin{array}{cccccccc}
1 & 0 & 1 & 1 & 1 & 0 & 0 & 1 \\
1 & 0 & 1 & 0 & 1 & 0 & 1 & 0 \\
\hline
1 & 0 & 1 & 0 & 1 & 0 & 0 & 0
\end{array}
$$

In this modification $0 \times 0 = 0$, $1 \times 0 = 0$, and $0 \times 1 = 0$, so only where both columns are 1 does the result become 1. Here, the trick is to treat each column on its own with no carry from previous columns, since this is a mask operation rather than an operation such as multiplication. Modification in this way is often called an AND operation. Only if the number in the input data *and* that in the mask are both 1 is the result a 1.

Another operation is the OR operation. In this operation, only if the number in the input data or that in the input mask (or both) were a 1 would the result be a 1. The effect of the OR operation on the same input data is

$$1 \quad 0 \quad 1 \quad 1 \quad 1 \quad 0 \quad 0 \quad 1$$
$$1 \quad 0 \quad 1 \quad 0 \quad 1 \quad 0 \quad 1 \quad 0$$

$$1 \quad 0 \quad 1 \quad 1 \quad 1 \quad 0 \quad 1 \quad 1$$

Finally, the last operation is known as exclusive OR or XOR for short. In this operation if the input data is a 1 or the input mask is a 1, but not both, then the result is a 1. The effect of the XOR operation on the same data is

$$1 \quad 0 \quad 1 \quad 1 \quad 1 \quad 0 \quad 0 \quad 1$$
$$1 \quad 0 \quad 1 \quad 0 \quad 1 \quad 0 \quad 1 \quad 0$$

$$0 \quad 0 \quad 0 \quad 1 \quad 0 \quad 0 \quad 1 \quad 1$$

These operations will be used again in later chapters.

▲

▼ Bits and bytes

You have probably come across the term "megabytes" before when specifying the size of the hard disk on your computer. Digital radio systems revolve around the transmission of bits and bytes. When a digital transmitter sends a signal, it formats it as a sequence of 1s and 0s and sends each one, one at a time. Each 1 or 0 is known as a *bit*. A collection of eight bits is known as a *byte*, although bytes are rarely used in communications texts. So how can 0s and 1s convey your message?

Well suppose that you actually wanted to send a short text message to your friend, say the word "hello." A very simple coding scheme might just take the position of each letter in the alphabet and turn it into a number. So "h" is the eighth letter in the alphabet, so "h" is replaced by 8; "e" is replaced by 5; "l" by 12; and "o" by 15. Now the message looks like 8, 5, 12, 12, 15. The next stage is to turn these numbers into binary numbers, that is numbers that only use the digits 1 and 0. With one digit, it would only be possible to send one of two letters, "a" or "b" by sending either a 0 or a 1. With two digits, the message could be 00, 01, 10, and 11, allowing the letters "a" to "d" to be sent. In fact, with two digits it is possible to send twice as many letters as with one digit. With three digits it is possible to send twice as many letters as with two digits, namely 8; four digits allows 16 letters to be sent; and five digits allows 32 letters, the whole alphabet plus a few spare. So the next stage is to replace each number with its binary equivalent using five digits. As it happens 8 gets replaced by 01000 in binary, 5 by 00101, 12 by 01100, and 15 by 01111. So the total message becomes 01000, 00101, 01100, 01100, 01111. The receiver does not need the commas because it knows that each set of five digits corresponds to one letter, but they make the message easier to read in this book.

In fact, there is a universal system of coding letters called ASCII that uses eight bits for each letter. This allows for all the lower and upper case letters, symbols such as exclamation marks, and a number of spare codes used for a range of purposes.

▲

In a random access message, eight bits are used, allowing one of 256 possible "letters" to be sent. The choice of eight bits is a compromise between using as few bits as possible in order to minimize collisions and using enough bits so that two users are unlikely to select the same "letter" at the same time.

The next thing that the network does to ease problems is to design into the mobiles the knowledge that their message might not get through because of a collision. So having sent a message, the mobile waits for a response to that message. If none is forthcoming, then the mobile assumes that its message collided with another mobile's and it needs to resend it. The worst possible thing would be for both mobiles to resend their messages immediately after realizing that they did not get through. In this case, both messages would collide again, and this could carry on forever. So the mobile waits for a random length of time (it does the equivalent of spinning a roulette wheel and waiting for as long as the number that comes up on the roulette wheel). It is unlikely (but not totally impossible) that the other mobile will have come up with the same number. Hopefully, the message will get through this time. The mobile will try 3 or 4 times before finally telling the user that the network is congested (normally this is indicated via an unhelpful, vague series of bleeps).

▼   **Random access channels and catastrophic failure**

Random access channels have a rather unhelpful property—if the number of mobiles attempting to send a message gets too high, nobody gets through at all. This can be imagined quite simply. The more mobiles that try to use the channel, the higher the chance that messages from two mobiles will collide. Eventually, there are so many collisions that no messages get through. The actions of the mobiles actually make it worse. When there is a collision, the mobiles keep resending their message until it gets through. Suddenly, the amount of traffic rises dramatically. At the point where the traffic level was just below the point of collapse there were a few mobiles sending a message mostly just once. As a few more mobiles attempt to access the network, now there is a slightly larger number of mobiles but each one is sending its messages over and over again rather than just once, dramatically increasing total traffic levels. In just a short period of time, the traffic level on the channel has risen dramatically. Theory predicts that this happens at approximately 35% capacity. That is, when there are so many messages that the channel is used for 35% of the time, the traffic will suddenly increase dramatically until the channel becomes unusable. It will not come back into use until most of the mobiles go away and the number of mobiles drops so low that they can get through. It is a little like an accident on the freeway. Once the tailbacks have built up, they do not go away until well after the rush hour, even if the accident is cleared quickly.

The network must make sure that the load on the random access channel never exceeds 35% if it does not want the whole cell to come to a standstill. To help it achieve this, it can stop certain mobiles sending a random access message. It can transmit a command that says "all mobiles whose telephone number ends in 1 (or any other number between 0 and 9) must not transmit a random access message until further notice." It can keep removing 10% of all mobiles from those allowed to make a random access message until the load comes under control. Each group of 10% of mobiles is known as an "access class." In addition to the 10 obvious classes, there are six special classes for the emergency services that cannot be turned off, leading to a total of 16 that is coded using four binary digits.

Access classes prevent catastrophic failure in a cell, but nevertheless, blocking 10% or more of the mobiles from making a call does nothing for customer satisfaction. If access class blocking needs to be used frequently it is an indication that the cell cannot provide sufficient capacity and should be replaced with two or more smaller cells, thus increasing the network capacity.

▲

With the use of repetition, if needed, the mobile's message has hopefully gotten through. But all that the network has received is a random number—not particularly helpful. Nevertheless, the network has learnt something. It knows that there is a mobile in that cell that wishes to communicate with the network. The cell sends back a message to the mobile. The message has the form "would the mobile that just transmitted random number 00110011 please send some more details on channel 23, which I am setting aside for its sole use." The mobile listens to all these messages until it finds one with the same random number, takes note of the channel it has been given, and sends a more detailed message on this channel. Because the channel has been dedicated to it alone, it does not have to worry about collisions and can send all the information needed.

Well not quite. There is one unusual situation that could occur. Two mobiles could send the same random number at the same time. Normally, both messages would collide and the mobiles would need to repeat their transmission. But say one mobile was very close to the transmitter and the other right on the edge of the cell. The signal from the mobile close to the transmitter would be so much stronger than the other mobile's signal that it might be possible to decode it at the cell site. This is a little like one person talking over another by talking more loudly. Although they interfere, it is possible to determine what the louder person is saying but not what the quieter one is saying. So the network receives one message. However, neither the network nor the two mobiles know what has happened. Next the network assigns a channel to the mobile sending the random number. But both mobiles will receive this message and think that the channel is for them. The means whereby this is overcome is described subsequently.

When the mobile transmits on the channel it sends a message of the form "I am mobile telephone number 0468 123456. I would like to make a telephone call to mobile 0860 123456." The network sends back a message saying "OK mobile 0468 123456, please wait while you are connected." At this point, the mobile on the edge of the cell will realize that the channel is actually for a different mobile, since the mobile phone number in the message sent by the network will not be its own, and will go back to the random access channel to send its random request all over again.

All this might seem a bit of a divergence. The section started by talking about how a mobile responds to receiving a paging message from the network. Well, it actually responds in just the same way as if it wanted to make a call; it also sends a random access message on the random access channel. When it is allocated a channel it says to the network "I am mobile telephone number 0468 123456. I have made this random access because you told me there was a call waiting for me." The network then connects the incoming call with the channel on which the mobile is waiting.

Given all the difficulties with random access, it would be reasonable to ask why, when a paging message is sent, a channel is not reserved for the mobile since the network is eventually going to reserve a channel for it anyway. The reason why this does not happen is the perennial problem—a lack of radio spectrum. The network only knows where the mobile is to the level of a location area. This might contain, say, 10 cells. The network would need to reserve a channel in each of these 10 cells in case the mobile was in the cell. Only when the mobile accessed the channel in one of the cells could the other nine channels be released for other mobiles to use. Given that there could be paging messages for hundreds of mobiles at the same time, and with collisions it might take a few seconds before the mobile was found, this could result in most of the capacity of the network being reserved and none left for calls to actually take place. Hence, it was decided that the mobiles should not have channels reserved for them.

So to recap, the network sends a paging message for the mobile in all the cells in the location area where the mobile is known to be. The mobile listens out for such messages and responds by sending a message containing a random number on the random access channel. It may need to repeat this message if it gets no response from the network. Eventually, it will get a message back containing the same random number and details of which channel to use. The mobile can then transmit on that channel, telling the network that it has been told that there is a call for it. The network can then connect the incoming call with the mobile. The amazing thing is that all this takes place in the space of a couple of seconds!

Problem 5.1

In which cells are paging messages sent?

Problem 5.2

How many different "letters" can be represented with a nine-bit message?

Problem 5.3

What is the total sequence of messages making up the establishment of a call to a mobile?

Problem 5.4

Why can the random access channel suddenly become overloaded, and what can be done about it?

Problem 5.5

Why does the network echo a mobile's number back to it in the message it sends back on the dedicated channel provided as a result of a random access attempt?

6

Hand-off

6.1 The mobile in a call

After having overcome all the hurdles in the last chapter, the mobile now has a dedicated channel all for its own use. It does not have to worry about interference from other mobiles, the network ensures that no other mobiles are allowed to use the same channel. All would be fine if the mobile stayed still. Unfortunately, there is little point in a mobile that is not mobile! The movement of the mobile causes some additional problems that must be accommodated.

6.2 Handing the mobile to another cell

As the mobile moves around, it is quite possible that it approaches the edge of the cell. This is the point at which the radio signal is too weak for the mobile to be able to talk to the cell. There is probably another cell neighboring onto this cell, but this neighbor cell uses a different frequency and knows nothing about the mobile. What is needed is a mechanism to allow the mobile to move from the current cell to the neighbor cell while the call continues. This is known as hand-off.

Hand-off is a major task for the network. It has to decide when to hand the mobile over and to which cell. It has to make sure the mobile actually managed to hand over correctly and it has to reroute the call to the new cell. And it has to do it so quickly that there is not a noticeable break in the speech signal[1]. By far the biggest problem in all this is deciding when and to which cell the mobile should be handed over. Hand-off is a major operation. It uses a up a lot of radio spectrum and can run the risk that the mobile gets lost in the process.

In the GSM system, the information on which the hand-off decision is to be based comes from the mobile, but the decision is made by the network. The mobile is well placed to measure the signal level received from surrounding cells and report this back to the network. However, it does not know whether there are any free channels in the surrounding cells, so the decision is best made by the network.

When a mobile is in a call, as well as listening to the channel on which the incoming call is being received, it also listens to signals transmitted by surrounding base stations. How it manages to do two things at the same time is a topic for latter chapters. It measures how strong (how "loud") the signal is from each of the surrounding base stations and continually feeds this information back to the cell in which it is located. It also feeds back information on the strength of the signal in its current cell. Simply, if the signal in the current cell is getting dangerously weak, the network looks at the signal levels from the other cells and spots one with an acceptable level. It then tells this new cell to reserve a channel for the mobile. When this channel is ready, the current cell sends a message to the mobile saying "switch to channel 45 now." The mobile makes this change and sends a message on channel 45 saying "I am here, is everything O.K?"

---

1. Until now the main discussion concerned signaling rather than speech. The speech channel will be reviewed in later chapters.

Once the network has this message, it shuts down the channel in the old cell and the mobile has been successfully handed over to the new cell.

That is mostly all there is to hand-off, although as will be seen later, the need to listen to other cells and to send information back to the current cell has major design implications. There is a slightly more unusual use of hand-off that is worth explaining. Say that there is a particularly heavy load in one cell. Perhaps there is a football match just finished and everyone is phoning home. Some mobiles in the cell may be in a position where they can receive adequate signal levels from the current cell and from another cell. The network might then decide to hand these users over to the neighboring cell even though their signal level was not low in order to spread the load over the surrounding cells and reduce the probability of congestion. This sort of clever tactic to reduce congestion is only possible if the network, rather than the mobile, makes the decisions to hand over to a different cell.

6.3 Cell selection in idle mode

Similar to hand-off is the situation of a mobile in "idle mode." As the mobile moves, it needs to decide to which cells it should listen in order to check for paging messages. In this case, it cannot rely on the network to tell it. Because the mobile is not in conversation with the network, the network has no idea that the mobile is coming to the edge of a cell. The mobile needs to do it all alone. As with hand-off, it listens to the signal strength from its own cell and surrounding cells. When the signal level from its own cell gets low it picks the surrounding cell with the highest signal level. It listens to the transmitted information in this cell and, if it discovers that it has entered a new location area, sends a message to the network saying that it is in a new location area.

It is now easy to explain how this message is sent. Basically, the mobile makes a random access in just the same way as if it were responding to a paging message or if the user had pressed the send button. When it gets a channel it says to the network "just letting you know I am now in this location area." The network sends back an acknowledgment and the conversation is over. The mobile returns to monitoring the paging channel.

Problem 6.1

Where is the hand-off decision made?

Problem 6.2

Why is the hand-off decision made there?

7

Network Management, Roaming, and Authentication

7.1 Network management

As was already mentioned briefly, as well as the three key components of base stations, mobiles, and switching centers, networks also have management centers that have a wide range of functions including:

- Monitoring the correct working of the network and raising alarms when failures occur;

- Delivering statistics about the network so that the need for additional resources can be monitored;

- Changing network configuration;

- Allowing the entry of new subscribers onto the network database and the deletion of old ones;

- Generating bills for users;

- Monitoring the network for fraudulent use;

- Providing customer care by monitoring usage, providing new features, and managing supplementary services.

Each of these is described in more detail in the following subsections.

7.1.1   Monitoring the correct working of the network

Any part of the network can fail at any point. Base stations could malfunction, the links between the base stations could be cut, or the switch could crash. The network operator needs to know about these problems as soon as possible so that they can take remedial action before subscribers become too annoyed about the loss of service. By having what amounts to a central computer that is continually communicating with each of the network components, one person can monitor the entire network and be informed immediately of any failures. In GSM, each of the base stations reports back on its correct operation to the BSC that controls it. The BSC is then responsible for reporting back to the network management system both about its own functionality and the functionality of all the BTSs under its control. This reporting takes place along the same cable as is being used to carry the voice conversations and the format used is standardized so that BSCs from different manufacturers can be connected back to the same management center.

7.1.2   Delivering network statistics

Running a mobile network is a careful balance between providing sufficient capacity so that the blocking targets can be met or exceeded, without providing so much capacity that the network costs rise. The point at which this balance occurs is continually changing as more subscribers come onto the network and as the calling patterns of subscribers change. The operator needs some key statistics so that he can check that his targets are being achieved and can receive advance warning of problems that may occur in the future. Statistics of interest include the percentage of time for which all the channels on a base station are in use (which is generally a good indication of blocking) and also statistics on dropped calls and voice quality levels reported by the network.

In fact, networks are rather good at generating these sort of statistics, so good that a day's worth of statistics from a typical network could easily be a few gigabytes of information. The real challenge to the operator is to filter this information so that only the statistics of interest are presented. To do this, most operators have developed complex software packages that look for deviations from particular targets and raise alarms. For example, if the number of minutes that all the channels on a base station are in use exceeds a certain level, this can be signaled to the operator, who can install additional cards in the base station to increase the capacity.

7.1.3 Changing network configuration

As networks evolve, it is often necessary to change the parameters of base stations, particularly with mobile networks. For example, it may be decided to split a cell into a number of smaller cells. This will mean that the frequencies used in the cells neighboring onto the split cell will typically need to be changed. Rather than physically having to go to the base station and change a switch in order to move to a different frequency, it is much easier to do this remotely. Typically, the operator will plan the new frequency arrangement and build a file showing all the changes that are required to the network. He can then arrange for this file to be processed at a quiet time (e.g., 1 a.m.) when all the changes can be made simultaneously to all the affected base stations, allowing the new frequency plan to come into place. Some GSM operators estimate that changes are made to their network parameters at least once a week. Clearly, having this centralized provides for a major saving in time and effort.

7.1.4 Adding and deleting subscribers

In order to use a GSM network you must be registered as a valid subscriber. The network must be informed about your phone number, your secret codes that will be used for security, and your particular service profile (for example, whether you are entitled to service when you roam to a different country). This information is stored in the HLR, but the operator needs some user-friendly means of getting it in there, retrieving and modifying it, and deleting it when the subscriber decides to leave the network. To do this they typically use a customer care package, which is essentially a database system running on a computer that is able to talk to the HLR. In some cases, this is a distributed network, with a terminal in each of the shops where users can sign up to the network so that the user details can be entered at the point of sale. There is little that is clever or new here, this is simply database management.

7.1.5   Generating bills

Subscribers should receive bills each month that need to be generated automatically by the network. Within the GSM specification is an item called a billing record. For every single call made, a billing record, which is an electronic receipt, is generated. This contains details of the person who made the call, the destination of the call, the length of the call, the location from where the call was made, and details of any special services (such as conference call) invoked during the call. The billing record is generated within the MSC where the call is generated. The MSC then uses the number of the originating subscriber (which tells the MSC where their home network is) to forward the billing record to a device normally located alongside the HLR, known as a billing system.

In essence, the billing system collects all the billing records and then at the end of each month generates a bill for each of the users, prints it out, places it in an envelope, and sends it to the customer. Like customer care systems, billing systems are little more than large database packages. However, they can be very costly because they need substantial flexibility to allow for different tariff packages and to charge calls differently depending on the time of day, for example. Because the bill is one of the few things that the customer gets from the network, often much time and care is spent ensuring that it is correct.

Billing systems can do more than this. They can detect when a subscriber's calling pattern changes which might indicate that the subscriber perhaps has new employment and is now using the phone more, and they can send a letter to the subscriber suggesting that they change to a different tariff. Like the network management system, they can also provide valuable statistics such as average call length, which can be useful when designing or modifying the network.

7.1.6 Monitoring fraud

Fraud is a big problem to mobile operators. Stolen phones can be used to create large bills for which the operator will get no revenue. Fraudsters think of a range of complex and clever schemes whereby they can make calls without paying for them, often involving complex call diversion routines. It is in the interests of the network operator to detect this fraud as early as possible and to remove the offending mobile from the network (by removing its right to service from the HLR). Basically, fraud packages monitor any changes in phone usage. If a phone that is hardly ever used for international calls is suddenly used nonstop for international calls, it is likely that this phone has been stolen and a fraud package will immediately remove this phone from the network. The difficulty with a fraud package is finding a balance between rapidly disconnecting stolen mobiles while allowing users to vary the traffic they generate from month to month. Fraud packages are getting more sophisticated but so are fraudsters, and this is likely to be a continuing battle for mobile operators.

7.1.7 Providing customer care

Mobile operators are often looking for ways to provide better service to their customers. This might include giving them information about football results over the GSM messaging service, updating them on major share movements, or providing a whole range of facilities. Alongside all the other computers needs to sit a flexible system able to provide these sort of services. To some extent, such a system is no longer part of the mobile radio network but forms the bridge between the network and external systems or sources of data. Typically, for each new service provided, special software will be required and it may be necessary to even update the computer system.

7.2   International roaming

GSM is one of the few mobile radio systems that allows users to go to a different country and still make and receive phone calls. This is a feature that is much valued by some users but requires increased complexity in the network.

When a phone is turned on in a different country, it has to do a lot of detective work. It needs to determine which networks it can receive and which of these it is allowed to use. First, it listens to all the channels, looking out for a channel that provides network information (the precise way that it does this will be explained in Chapter 11). When it finds such a channel, it listens to the information on this channel and finds out the name of the network. Within the mobile is a list of networks that it is allowed to use. It compares the received network with this list and, if they match, sends a message to the network saying "Can I please have permission to use your network."

The network to which the mobile has roamed does not know anything about the mobile. It needs to find out whether this is a valid mobile and how it should treat it. Using the first set of digits in the mobile number for call routing it can send a message over the PSTN to the network that the mobile normally uses, the so-called *home network*. This message ask the home network to send it all the relevant information about this mobile, including details of whether the mobile is allowed to roam internationally, whether the mobile is stolen, and details of security features that the mobile is using. The message also tells the home network to make a note in the HLR that this mobile is now in a different network and to note the details of how to send a call to that network.

Once the visited network has received a response from the home network and is satisfied that the mobile is valid, it sends a message to the mobile saying that it is able to get service in this network. At this point, the name of the network appears on the mobile screen and the user is able to make and receive calls. The whole process typically takes around 30 sec, although some mobiles are faster than others.

If the visited network does not receive a satisfactory response from the home network, then the mobile is refused service. This might happen if the user did not have a subscription for international roaming or if they had not paid their bills recently or if the home network had not set up an agreement with the visited network to allow roaming between the two networks.

7.3 Authentication and ciphering

Security is now a major concern with mobile networks. Security comes in two guises: making sure that nobody can overhear your conversation, which is known as *ciphering*; and making sure that the mobile is who it claims to be in order to prevent another mobile's calls from appearing on your bill, which is known as *mobile authentication*.

Before finding out more about security, it is necessary to understand what a SIM card is. SIM actually stands for *subscriber identity module*, but, like GSM it is one of those abbreviations that has become understood without people remembering for what the acronym actually stands. A SIM card is a credit card-sized piece of circuitry that is placed in the GSM phone. The SIM card contains all the personal information, including your number, your secret number used for security (explained subsequently), and other information such as the short-dial telephone numbers you have programmed. The advantage of a SIM card is that should your mobile malfunction, you can simply take the card out of the mobile, put it in a different mobile (which hopefully your kind retailer will lend you), and immediately be able to use the phone to send and receive calls.

Another use for the SIM cards occurs in the United States where some of the *personal communications services* (PCS) operators are offering a GSM service. However, this is in a different frequency band from that in use in the rest of the world; hence, "normal" GSM mobiles will not work in the United States. However, if you take your SIM card to the United States you can hire an American GSM mobile, put your SIM card in it, and make and receive calls, which will be billed in just the same manner as if it was your own mobile. This is known as *SIM roaming*.

Not all SIM cards are the size of a credit card. Indeed, the SIM card is actually only the size of a postage stamp, but is typically packaged in a credit card to make it easier to handle. Mobiles that advertise themselves as having a "full-sized SIM" have got their SIM card packaged in a credit card sized bit of plastic. Other mobiles have just got the postage stamp SIM.

Now back to security. There are many different ways of ensuring security. This section uses the GSM system as an example. Central to security in the GSM system is a user's secret number. Each user is given a unique number that is stored in their SIM card and in the network (in the AuC—see Figure 2.4). This number is very carefully guarded because if it became known by someone else the system of security would break down. The number is never sent from the AuC, so it cannot be overheard in transit. Similarly, the number never leaves the SIM. In particular, the number should never be sent by radio since anyone could listen in and receive the number, put it in their mobile, and pretend to be someone else's mobile. This was the mechanism whereby some of the early cellular phones were "cloned."

This is what happens when a mobile starts using a channel. The network asks the AuC for a random number and a response. The AuC then performs a special operation (called a "security algorithm") where the random number and the secret number were the two inputs and the response is the output. This is shown in Figure 7.1.

The network then sends the random number to the mobile. The mobile knows the special operation performed by the network and so can put the random number and its copy of the secret number into the

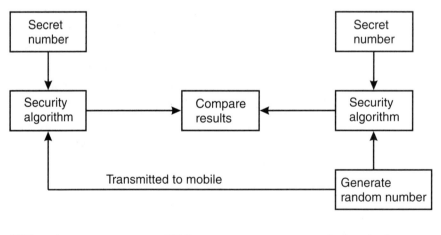

Figure 7.1 Authentication.

operation. The result of the operation is then sent back to the network. If the result received from the mobile is the same as the result received from the AuC, then the mobile is judged to be who it says it is.

In order to be highly secret, the special operation must be designed so that even if an eavesdropper knows the random number, the special operation, and the response, they are not able to work backward to deduce the secret number. For example, a multiplication operation would be no good because the eavesdropper would only need to do a division to get back to the secret number. Security experts have worked hard to ensure that the special operation is such that it is almost impossible to work out the secret number. For extra security, the special operation itself is kept secret, so it is not possible to explain it in more detail here. However, the security of the systems rests upon the secret key remaining secret and not on the special algorithm remaining secret.

So now the mobile is authenticated. The next stage is to encipher the call. First, a few words about how ciphering works. Both sides agree on a mask, which is a sequence of 0s and 1s. The transmitter then multiplies the data to be sent by the mask and sends the result. The receiver multiplies the received signal by the same mask to get the original signal. This can be best seen by an example. Remember from Section 5.3 the binary XOR operation where if you multiply a 0 by a 0, you get 0. If you multiply 0 by 1 you get a 1, while if you multiply a 1 by a 1 you get 0.

Now assume that the mask (the "multiplied by" number) is going to be eight-bits long (in practice much longer masks are used, but eight bits allows a workable example) and that both sides, by some mechanism to be explained, agree that the mask will be 11011100. It does not matter what the mask actually is as long as both sides agree on the same thing. Now look at what happens when the input data is multiplied by the mask, transmitter, and multiplied again, as shown in Table 7.1.

To follow this example, take any column, say the first one. The input data is 0 and the mask is 1. The XOR operation with 0 and 1 returns a result of 1, which is the transmitted data. At the receiver, the transmitted data is multiplied by the mask, so the transmitted 1 is XORed with the 1 of the mask. The XOR operation with a 1 and a 1 returns a 0, which is the received data. Hence, the received data is the same as the transmitted data. Multiplying twice by the same mask results in receiving the original data. The same procedure is followed for all the other columns.

Table 7.1
Multiplication by a Mask

Input data	0	0	1	0	1	1	0	1	1	1	0	0	1	0	1	0
Mask	1	1	0	1	1	1	0	0	1	1	0	1	1	1	0	0
Transmitted data	1	1	1	1	0	0	0	1	0	0	0	1	0	1	1	0
Mask	1	1	0	1	1	1	0	0	1	1	0	1	1	1	0	0
Received data	0	0	1	0	1	1	0	1	1	1	0	0	1	0	1	0

Like the secret number, the mask cannot be transmitted over the radio because otherwise someone could hear it and use the same mask to decode the signal. Instead, both the mobile and the AuC use the same random number that was already sent, the secret number, and a different special operation to generate the mask. As long as they use the same secret number and the same special operation, they will come up with the same mask. They can then start to encipher their conversation.

7.4   Call routing

Roaming leads to one particular problem when it comes to routing the calls. Say John is a U.S. subscriber but has roamed to France and that Michel, a cellular subscriber in France wishes to call John. Michel dials John's number, which is, of course, a U.S. number. This call request passes into the PSTN, which routes the call to the United States. Once it gets there, the HLR in the U.S. notes that John is in France and routes the call on to France where it finally reaches John. So despite the fact that John and Michel are in the same country, the call ends up going via the United States. This is shown in Figure 7.2. The problem is known as *tromboning* since the path the call follows looks a little like the long arm of a trombone. The basic problem is that the PSTN and gateway switches in combination have very little intelligence and are not able to work out that this call comprises a redundant international segment that could be collapsed.

There could be a number of solutions to this problem. The most obvious one is that instead of routing the call to the United States, the PSTN could just send a message asking where the subscriber actually is and then

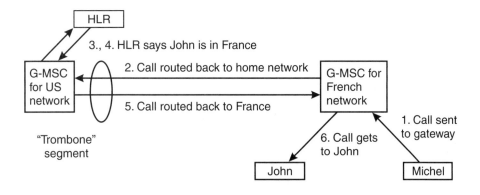

Figure 7.2 A trombone.

routes the call based on the answer. Unfortunately, the PSTN is not designed to work in this way, although future versions of GSM will have this capability so that, at least, if Michel is calling from a GSM mobile, his call will reach John directly rather than via the United States.

There is another issue associated with paying for international calls. Now, say, Peter, based in the United States, wants to call John. He has no idea that John is actually in France and just dials the number as normal. What happens is that an international call is set up, costing more than a national call. It would not be fair for Peter to pay the higher rate with no warning—he might have chosen not to make the call if he knew that it was going to be charged at a higher rate. Peter could be warned by a voice message that the call was going to be at a higher rate and that he should hang up now if he did not want to pay this rate. However, this would tell Peter that John was not in the country, and there are some mobile phone users who would rather that their callers did not know they were out of the country.

There is only one solution in this case. Peter will pay what he would have paid if John was in the country, while John will pay whatever extra cost there is as a result of the international part of the call. Because some users might not want to pay for incoming calls while they are in a different country, GSM phones have a "bar incoming international calls" capability that means Peter would get the same message as if John's mobile had been turned off.

Problem 7.1

What is the effect of multiplying the data 01101010 by the mask 00110011?

Problem 7.2

How is the mask derived in GSM?

Problem 7.3

Under what situations will a trombone occur?

Problem 7.4

How do users determine their secret numbers?

Problem 7.5

How does a visited network know which is a mobile's home network?

Part III

Transmitting the Signal

I N PARTS ONE AND TWO of this book overall system design and some of the higher level functionality were explained. The rest of the book focuses on the design of cellular and wireless systems in more detail. Until now, it was just stated that the transmitter sends a signal to the mobile. The process of reliably transmitting a radio signal is very involved and forms the content of this third part of the book.

This section revolves around the block diagram of a radio system, which will be presented in Chapter 9. However, before the reason for some of the blocks can make any sense it is important to understand what happens to a radio signal in the process of being transmitted. From transmittance, the signal is said to propagate to the receiver; hence, the study of what happens to the signal during transmission is known as radio propagation.

8

Radio Propagation

8.1 Introduction

Propagation is the term used by engineers to describe the radio signal traveling from one point to another. The signal is said to propagate from the transmitter to the receiver. Radio propagation plays a key role in all radio systems. The propagation that can be achieved limits the range of the radio signal and hence the size of the cells. Propagation phenomena mean that some subscribers in a cell are unable to obtain a satisfactory signal. Propagation causes some frequencies to be more desirable than others. Understanding propagation is key to understanding radio technology.

First consider an analogy. Radio propagation could be compared to sound propagation. If you attempt to shout across a field to someone on the other side, they might be able to hear you some of the time. But sometimes the wind will be in the wrong direction and on other occasions a passing tractor will drown out the sound of your voice. To make sure that you are heard, you will shout as loudly as you can and may repeat what you say to ensure that errors do not creep into you message. If you consider radio propagation akin to shouting across a field, then you will start to understand the problems involved.

If you transmit a radio signal and look at the strength of the received signal as the receiver moves around a town, the resulting trace of signal strength will look extremely complicated, as shown in Figure 8.1. There will be lots of downward and upward spikes, occasional longer term fluctuations, and no obvious reason for all these variations.

Much research has been performed to try to rationalize the signal strength that is received. To simplify the problem and to help predict the received signal, mobile radio planners consider that the path loss experienced when transmitting a signal through the channel is composed of three distinct phenomena:

- Distance-related attenuation;

- Slow fading;

- Fast fading.

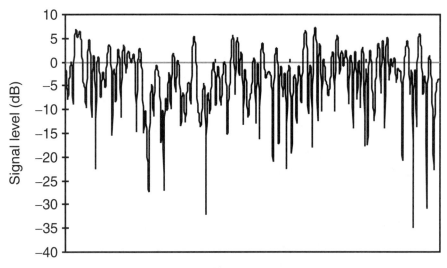

Figure 8.1 Rayleigh fading waveform for a mobile moving at walking speed over 1 sec.

8.1.1 Distance-related attenuation

Distance-related attenuation simply expresses the fact that as the distance from the base station increases, the signal strength decreases. This is entirely consistent with everyday experience where the further you move from someone who is talking, the weaker the signal. Alternatively, imagine throwing a stone into a pond. Ripples spread out in a circle from the point where the stone entered the water. As the ripples get further away they get progressively weaker until they finally disappear.

In the case of a radio signal, the drop in signal strength is caused by the fact that the signal spreads out from the transmitter as if on the surface of a sphere. This is the so-called "free space propagation." The area of the surface is proportional to the radius squared, and hence the signal strength is inversely proportional to the distance squared from the transmitter to the receiver. This is written mathematically as $1/d^2$, where d is the distance from the transmitter. So at a distance of 4m from the transmitter, the signal strength is equal to one-sixteenth of the strength at 1m away; and at 100m, the signal strength is only one-ten thousandth of the strength at 1m away.

Measurements of mobile radio channels have found that, in practice, the signal strength decreases more quickly than $1/d^2$. A typical value often used in predicting mobile radio propagation is $1/d^4$; in this case the signal at 100m away is only one-hundred millionth of the signal at 1m. The reason for this more rapid reduction in signal is that the radio signals are absorbed by vegetation and buildings.

▼ Coping with large numbers in mobile radio

The first pass look at signal propagation has shown that the signal received by the mobile might be a very small fraction of that transmitted. If the propagation factor was $1/d^4$ and the receiver was 4 km from the transmitter, then the signal received would be $1/4000^4$ weaker than the signal at 1m. Now $1/4000^4$ equals 0.0000000000000039. Writing down numbers like this is cumbersome and it is very easy to make a mistake. The solution to this problem came from Alexander Graham Bell himself. He said that life would be much easier if you used the logarithm of the number rather than the number itself.

The logarithm of a number is defined as the power that you need to raise 10 by in order to get to the number. Now $100 = 10^2$, so the logarithm of $100 = 2$. $1000 = 10^3$, so the logarithm of $1000 = 3$. Some more simple logarithms are shown in Table 8.1.

To find out the logarithm of a number, just type the number into your scientific calculator and press the button marked "log." If we take our earlier number of 0.0000000000000039 and take the log, the answer is −14.4, which is much simpler to handle. Numbers in this format were called Bels, after their inventor, so this number would be written −14.4B. However, Bels go almost too far the other way in that quite large differences between numbers result in small numbers of Bels. To make life even more convenient, it was suggested that this number be multiplied by 10 and called a decibel. So now our number becomes −144 dB.

Decibels are the language of radio engineering. If a signal is 10 dB stronger than another, it is 10 times as strong. If it is 20 dB stronger, it is 100 times as strong. Engineers always remember that 3 dB stronger corresponds to twice as strong.

Decibels just represent a ratio, that is, the difference in signal strength at two different points. If you want to measure actual signal strength, you have to make the signal level relative to something. Typically, engineers use the signal strength from 1 W (Watt) of power as a reference level, and

Table 8.1
Some Simple Logarithms

Number	Logarithm
10,000	4
1,000	3
100	2
10	1
0	0
0.1	−1
0.01	−2
0.001	−3
0.0001	−4

hence a signal measured in decibels relative to Watts (dBW) is the signal power relative to 1W. So if the power was 20 dBW, this would correspond to 100W.

▲

8.1.2 Slow fading

Slow fading is a mobile radio phenomena caused by the mobile passing behind a building. During the period the mobile is behind the building, the signal received will be reduced. Driving along a road, the mobile will pass behind a sequence of buildings, causing the signal to reduce in strength, or *fade*, on a relatively slow basis (compared to fast fading as explained next).

8.1.3 Fast fading

Fast fading is another mobile radio phenomena caused by the signal arriving at the receiver via a number of paths. Imagine that two signals are received at the mobile. One passes directly from the base station to the mobile via a *line-of-sight* (LOS) path. The other is reflected off a building behind the mobile and back into the mobile antenna. This is shown in Figure 8.2 and is often called *multipath propagation*. The mobile then sees a signal that is the composite of these two signals. The reflected signal has traveled a slightly longer path than the direct signal, and hence it is delayed slightly compared to the direct signal. The result of this delay is that the sine wave of the reflected signal will be at a different part of its cycle than that of the transmitted signal. Sine waves in different parts of

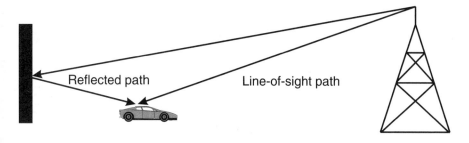

Figure 8.2 Multipath propagation.

their cycle are said to have a different phase. The size of the phase difference is related to the difference in distance divided by the speed of light (giving the delay on the signal) multiplied by the frequency of transmission.

This can be better understood by an example. Suppose that the reflected signal travels an additional 10m before arriving at the mobile. Light travels at 3×10^8 m/s, so it will take $10/3 \times 10^8$ (i.e., 3.3×10^{-8}) sec to travel this additional distance—a delay of around 33 ns (remember from Section 3.2 that a ns is a nanosecond, which is one-thousand millionth of a second or, in mathematical notation, 33 ns = 33×10^{-9} sec). If the frequency of transmission were 3 GHz, then one cycle of the carrier wave at this frequency takes 0.33 ns. Hence the reflected wave will be delayed by 100 cycles of the carrier wave. Every additional 10-cm distance will, in fact, delay the reflected wave by a further cycle of the carrier wave.

When the reflected wave is delayed by an exact multiple of a cycle of the carrier wave the two received signals are said to be "in phase." The receiver effectively adds the two signals together, such that if the reflected wave was of the same strength as the carrier wave, the total received signal has twice the strength of the direct signal. When the reflected wave is delayed by an exact multiple plus exactly a further half a cycle the two signals are said to be in anti-phase. That is, they are exactly opposite to each other. In this case, they cancel each other exactly. This is shown in Figure 8.3.

In these figures, the first shows a continuous sine wave—effectively the LOS path. The second shows a reflected path of the same signal strength (typically reflected paths are actually of a lower signal strength). This reflected path starts by being in phase but gradually changes to being anti-phase. The third figure shows the result of adding the two signals together. At first the resultant signal is twice as strong as the LOS path; however, as the two signals become increasingly out of phase, the resultant signal is gradually reduced to nothing.

That is to say, at a certain point the mobile will lose all received signal. By moving a distance of only half a wavelength then, some 5 cm, the mobile will move from a position where the signal strength is doubled to one where there is no received signal. This phenomena repeats continuously as the mobile moves; hence the signal fades rapidly, giving the name "fast fading" to this effect.

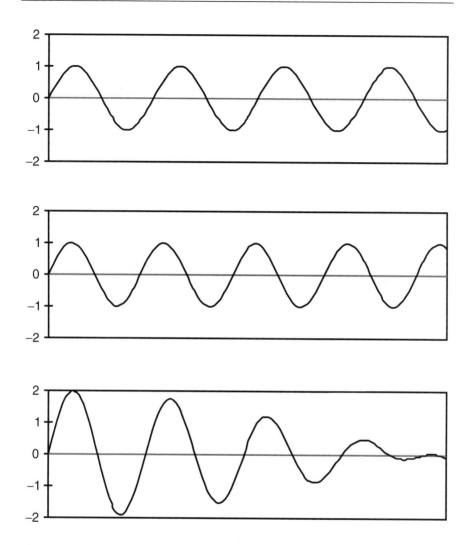

Figure 8.3 Addition of in-phase and anti-phase waveforms.

Of course, in real life, there are many more reflected signal paths, and the reflected signals are not all of equal strength, so the prospect of an exact cancellation is somewhat reduced. Nevertheless, fading is a severe problem. Figure 8.1 shows a typical fading waveform. This waveform is often termed Rayleigh fading after the mathematician who developed the statistics that can be used to describe such a waveform. It can be seen that in a period of 1 sec a number of fades, some as deep as almost 40 dB (that is the signal falls to one-ten thousandth of the level it would normally be), are experienced.

There is a further difficulty that may be caused by fast fading. Imagine now that instead of the reflection coming from a nearby building, it comes from some faraway mountain. The delay of the reflected signal may now be quite large. If this delay is greater than the time taken to transmit a bit of information, then when the reflected signal finally arrives it is carrying different information to the direct signal. The result of this is that the previous bit transmitted interferes with the current bit, causing a phenomena known as *intersymbol interference* (ISI). This problem is discussed in more detail in the next section.

8.2   Wideband channels

▼   Data rates, bits, symbols, and channel bandwidths

In the next chapter the conversion of speech into a digital waveform will be discussed in more detail. Suffice it to say at the moment that the result of the conversion of speech into digits is around 25,000 1s or 0s per second. Each of these 1s or 0s is known as a *bit*[1]. In a typical radio system, each bit is transmitted, one at a time. The radio system is said to have a data rate of 25k bits per second or 25 Kbps. Now, if you want to send 25 Kbps of data it so happens that you need 25 kHz of radio frequency. An analogy to this can be found with gramophone records (for those who still remember them). If you play a 45 record at 33 rpm, then it will take longer to play. The data rate being read from the record is reduced because there is a fixed amount of data on the record but it is taking longer to play it. If you listen to this record you will find that the sound has gotten deeper (singers' voices will sound deeper, or lower in tone). There is less-high frequency sound. Hence, the sound channel would not need to be so wide to carry this information. Now speed the record up to 78 rpm. The data rate increases because the record takes less time to play. The sound also gets much "higher" (singers' voices become more high pitched) and hence a wider sound channel is needed to carry all the data.

---

1. The term "bit" is actually a shortened form of "binary digit."

The same is true with radio—the faster you want to send data, the wider the radio channel needed to do it. The width of the radio channel required is known as the *channel bandwidth*.

▲

As mentioned briefly Section 8.1, reflection can cause an ISI problem, which is problematic in mobile radio systems having transmission bandwidths greater than around 100 kHz. In order to understand whether the problem might occur, consider the following example.

Assume that the data rate is 100 kbps (broadly this means that the bandwidth of the transmitted signal will be around 100 kHz). The carrier frequency is not a relevant factor in calculating ISI. If the reflection is delayed by 1/100 kHz, namely 10 μs, then the reflected signal will arrive during the time that it takes for the next bit to arrive on the direct path, causing ISI. In 10 μs, radio signals travel 3 km. Hence, if the reflection occurs from an obstacle 1.5 km past the receiver, ISI will occur.

As the symbol rate increases, the distance required for ISI to occur reduces proportionally. With a bandwidth of 1 MHz the distance falls to 300m, and at 10 MHz the distance falls to only 30m.

ISI is a problem because when a signal arriving by one path carrying one bit and another signal arriving by a longer path still carrying a previous bit arrive at the same time at the receiver, what the receiver sees is the combination of both of these. The receiver cannot distinguish between the two signals and hence cannot correctly decode the transmitted signal. However, if the receiver knows a little about the reflections it is actually able to remove the ISI using a device known as an equalizer. This is briefly discussed in the next chapter.

8.3 Predicting cellular coverage

To minimize system cost and roll-out time, operators need to ensure that they use the fewest number of cell sites possible to provide the required coverage. The problems will be familiar to the cellular operators who expend considerable time and effort planning their networks to use the minimum number of base stations for the required coverage. Cellular operators invariably deploy computer-based planning tools to aid them in this task.

Planning tools are basically computer programs that use as an input a digital map of an area. Using algorithms derived from propagation rules, coverage areas for hypothetical base station sites can be predicted. Different base station locations can be rapidly tried until a near-optimal result is achieved. Planning tools are available for cellular planning from a wide variety of vendors covering different technologies and cell sizes.

8.4   Sectorization

If you look at most cellular antenna installations, especially in cities, you will see an array of antennas, typically constructed in a three-sided or six-sided arrangement. This is known as sectorization. In a sectorized cell, instead of using an antenna that radiates signal equally in all directions (known as an *omni-directional antenna*, or *omni*), antennas that only radiate narrow beams of 120 degrees (in a three-sided arrangement) or 60 degrees (in a six-sided arrangement) are used. Taking, say, the three-sided case, this has the effect of splitting the cell into three equal-sized pie-shaped slices as shown in Figure 8.4.

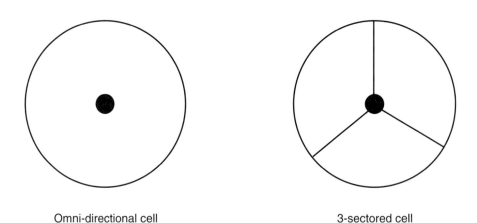

Omni-directional cell   3-sectored cell

Figure 8.4   An omni-directional and sectorized cell.

Sectorization brings a number of advantages. However, contrary to what might be imagined, by splitting a cell into three sectors it does not create three times the capacity. First we explain why.

In a three-sectored cell, the cell edge is still the same distance from the transmitter as in the omni case. Hence, the transmitter and mobiles still need to transmit with almost the same power as before (power levels can be reduced slightly due to the use of the directional antennas) and the distances until the signal can be reused are still similar. However, the same frequency cannot be used in each sector in the cell and hence more frequencies need to be found—three times as many in fact. Three times as many frequencies means only a third as many in each sector. Now each sector only has one-third of the number of subscribers as the omni (because it is only one-third of the size), and hence there is little change in the overall capacity.

In practice, some careful cell design and frequency planning can reduce the number of frequencies required to somewhat less than three times providing a small gain in capacity, but this gain is relatively minor compared to other savings that could be made. So sectorization does not significantly increase capacity (although, as will be seen later, there are some special systems where sectorization can actually bring capacity advantages).

Given the limited gains of sectorization, it is reasonable to ask why most cellular operators use a sectorized cell system in city areas. The reason is that sectorization allows for a greater path loss than a nonsectored approach because the gain of the sectored antennas provides the mobile with a stronger signal. This has the effect of increasing the range. Further, in city areas, sectorization prevents some multipath reflections that might occur if an omni-directional antenna was used, since signals are now sent in a narrower beam reducing the chance of reflections from, for example, behind the base station, and increasing signal strength. Finally, as explained previously, a small-capacity gain is achieved. Overall, for a small cost increase, improvements in capacity, range, and signal quality can be achieved by sectorization in this area.

However, in suburban and rural areas, sectorization is used less. This is because in these areas it may be that only one frequency is required to provide sufficient capacity. If sectorization was adopted, a different frequency would need to be deployed in each sector that would waste equipment, resulting in a significant cost increase.

8.5   Microcells and underlay/overlay networks

Chapter 3 showed that capacity is calculated in terms of Erlangs per cell. It also showed that in the case where there was not sufficient capacity in the network, the cells had to be made smaller. Basically, a cell has the same capacity regardless of the area it covers. As the area covered gets smaller, there will be fewer mobiles within the area and hence the cell will be able to provide the capacity required.

In fact, if you are asked to calculate the capacity of a cellular system with a given amount of radio spectrum, an easy answer is to say that the capacity is infinite given small enough cells. In practice, many practical issues means that really small cells are not possible. Chief amongst these is the cost. The cost of a cell does not depend (much) on the size of the cell, so as the cell size is reduced, more cells are required to cover the same area and the cost of the network rises. Other problems include the need to continually hand over subscribers moving through small cells and the increased network complexity as the number of cells increases.

Nevertheless, congestion in some networks, especially in city centers, has reached such a level that quite small cells are being considered. These cells are termed *microcells*. The definition of a microcell is one where the transmitter antenna is lower than the surrounding buildings. In this case, the radio signal is more-or-less restricted to the street where the transmitter is located. This typically results in cells some 500-m long in total (250m each side of the transmitter) and shaped in the same way as the street (i.e., a long cigar shape rather than the circle of the larger cells). If you take a crowded city center covered by, say, three normal cells and replace these by 30 microcells, your capacity will rise by a factor of around 10, as will your costs. Actually, neither of the following statements is entirely accurate.

■ The capacity will rise by a little less than 10 because there are fewer mobiles in each microcell; hence, fewer frequencies will be deployed and the trunking gain will be lower—but this can be ignored as relatively minor.

■ The cost will rise by a little less than 10 because microcell equipment tends to be less expensive and the transmitter sites used will not be on the top of the tallest buildings, and hence the site rental payments will be lower.

Microcells are often the only way to improve capacity in city centers. However, they do result in handover problems. First, more handovers are required because the cells are smaller. Second, handover decisions now need to be made much faster. In some cases, one microcell may cover one street and another a side street. When turning into the side street the mobile needs to be handed over very rapidly as the signal from the main street will fall to a level too low to be used for communications in around 10m. A car can cover 10m very quickly.

The typical solution to this is to leave the original large cells in place and to use these for handover. When the mobile detects that the signal strength is falling, it defaults back to the large cell while it takes the necessary time to find another small cell with sufficient signal strength. Such large cells also mean that the microcells do not need to cover the whole city, which can be extremely difficult to achieve, with the large cell providing coverage in any gaps.

Such an approach is known as a *hierarchical cell structure*. The large cells are known as *oversailing* or *umbrella* cells. The umbrella cells need to use a different frequency from the microcells to avoid interference. Typically, when the microcells are installed, most of the frequencies used by the oversailing cells are removed and shared amongst the microcells where they can be reused many times across the city. The oversailing cell keeps only a few frequencies to deal with handover and providing coverage in areas not covered by microcells. The MSC needs to make sure that handover decisions are taken to put mobiles on the microcells, where the main capacity resides, as often as possible.

The next chapter looks at the design of a generic radio system. This will be of use in understanding the advantages of particular radio systems. It will also show how radio designers overcome some of the propagation problems described in this chapter.

Problem 8.1

How much stronger is a signal that is 20 dB more than another signal? How much stronger is one that is 3 dB? How much stronger is one that is 23 dB?

Problem 8.2

What causes fast fading. If a radio signal has a wavelength of 15 cm, how far apart will the fast fades be?

Problem 8.3

Why would a cellular operator sectorize a cell?

Problem 8.4

How does ISI differ from fast fading?

Problem 8.5

Why does decreasing the cell size increase the capacity of a radio system?

9

Radio Systems

9.1 From speech to radio waves

In this chapter, the general design philosophies of radio systems are explained, concentrating on those parts of the radio system that are of relevance to cellular technology. Figure 9.1 shows a block diagram of a typical *digital* radio system. This is the process through which a user's speech goes before it can be heard by the listener. Note that the ordering of blocks is important: speech coding must be performed first, modulation last, and enciphering before error correction. In this section, the need for each of these blocks and the manner in which they are designed will be explained in more detail.

As you will see, each of these parts involves a number of stages, some of which are quite complicated. Despite the complexity, all these operations need to be performed very quickly. If they take more than around 10 ms, the delay on the speech will become annoying. Modern mobiles

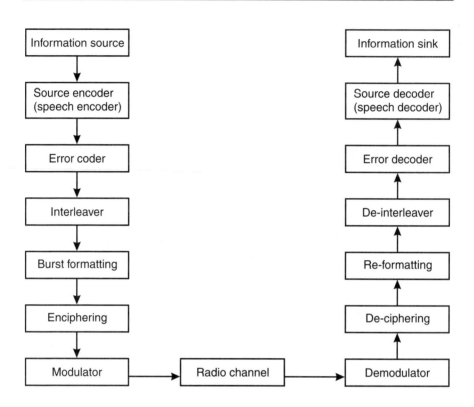

Figure 9.1 Block diagram of a radio system.

have almost as much processing power as a desktop computer to allow them to perform all these stages with minimal delay.

Each of the stages is now described in more detail, except ciphering, which has already been explained in Section 7.3.

9.2 Speech coding

In digital radio systems it is necessary to turn voice signals, which are analog, into a digital data stream. Analog signals are those that can have any strength whereas digital signals are restricted to having a strength equal to either a 0 or a 1. (If data is to be transmitted, rather than voice, then this stage is not required.) Speech coding is a highly complex topic on which thousands of researchers around the world are working and which is evolving daily. Here, a simplified explanation is provided because a full treatment is beyond the scope of this book.

The simplest speech coders are essentially analog-to-digital converters. The analog speech waveform is sampled periodically and the instantaneous voltage level associated with the speech converted to a digital level. A stylized diagram illustrating the process is given in Figure 9.2.

The amount of information transmitted and the quality of the speech are then dependent on two parameters. The first is how frequently the

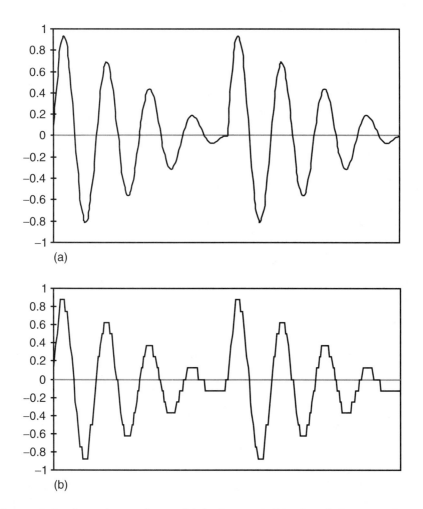

Figure 9.2   Speech waveforms (a) before and (b) after A-D conversion.

speech is to be sampled (the "sampling rate"), and the second is how many bits are used to describe the voltage level—the more the bits, the less the resulting waveform will look steplike. In technical terms, the difference between the digital waveform and the original analogvh waveform is known as the "quantization error." The trade-off is to provide a sufficiently high sampling rate and number of quantization bits to provide good voice quality while avoiding using so many that spectrum resources are wasted. In calculating the sampling rate, a theorem known as the Nyquist theorem can be used. This theorem states that if a waveform is sampled at twice the highest frequency in the waveform, then it can be correctly recreated. Voice is generally assumed not to contain any useful information above frequencies of 4 kHz. (This is not strictly true. Some sounds like "s" contain substantial energy above 4 kHz. The removal of these frequencies in telephone systems can make some sounds difficult to understand, particularly isolated letters.) Hence, a sampling rate of 8 kHz (i.e., 8,000 samples per second) is typically sufficient for an acceptable voice quality (at least equivalent to existing telephone systems).

There is an internationally agreed-upon standard for voice coding using 8-kHz sampling, known as *pulse code modulation* (PCM). This utilizes eight-bit sampling at 8 kHz, resulting in a bit rate of 64 Kbytes/s. This voice coding system is currently used in most fixed networks and digital switches (that is to say, that when you make a call using a fixed phone, it is highly likely that your voice is being digitized and transmitted using the PCM coding technique). Its voice quality is considered to be near-perfect.

From an information viewpoint, PCM is extremely inefficient because it takes no account of the fact that speech is made up of sine waves that are fairly predictable. Typically, over the period of one syllable, lasting perhaps 20 ms, the speech is unchanging and can be represented better by saying "the user is currently generating a sine wave of frequency 1,200 Hz and volume 6 dB." More generally, speech waveforms are smooth—changes from one syllable to another occur slowly and these facts can be used to reduce the bandwidth of the digital signal.

The simplest system that starts to use the fact that much of the voice waveform is predictable is known as *adaptive differential PCM* (ADPCM). This system still samples the speech waveform at 8 kHz, as with PCM. However, instead of quantifying the absolute signal level it quantizes the difference between the signal level at the previous sample and the current sample. Because speech changes slowly, this difference is typically small; hence, it can be accurately represented by fewer bits than the absolute level. This is the differential part of ADPCM. A further enhancement is to note that periods when speech waveforms are slowly changing tend to occur for some time, so during this time the size of the gap between each of the quantization levels can be reduced. This gap is increased as the waveform changes more quickly. Hence the quantization size becomes adaptive to the waveform.

ADPCM generally works well. However, because the differential is transmitted, if there is an error in receiving the differential, this will make the receiver represent a different point on the received waveform than that transmitted. This difference will continue while only differences are sent. Further, an error in reception could lead to the receiver changing the adaptive levels to different settings than the transmitter, resulting in further errors. These errors will propagate until the transmitter sends a reset-type signal, re-synchronizing the receiver. This needs to be per-formed regularly to prevent error propagation from becoming severe. Even with regular transmission of the reset signal, the effect of an error can be a short "glitch" in the speech. These types of error are the price paid for reducing the information used to transmit the signal.

Because ADPCM systems only use four bits to quantize the difference in the waveform, the bitstream generated is 32 Kbits/s, only half that of PCM. ADPCM is generally considered to provide quality virtually as good as PCM.

Information theory shows that it is possible to go much further. More optimal voice coders would model the vocal tract of the speaker based on their first few syllables and then send information on how this vocal tract was generating sound. The speech coders used by digital cellular systems follow this route to a greater or lesser extent. For example, GSM utilizes a coder type known as *regular pulse excited–long-term prediction* (RPE-LTP), where the LTP part sends some parameters showing what the vocal tract is doing and the RPE shows how it is generating sound ("being excited"). Such speech coders are extremely complex and beyond the scope of this book.

As far as cellular operators are concerned, the key parameters are the bandwidth required, which has direct implications on the number of subscribers that can be supported, and the voice quality. Some coders and the general perceptions about their quality are shown in Table 9.1.

For the WLL operator, voice quality can be particularly important. While users accept, unwillingly, that voice quality in a mobile radio system can be inferior to wireline quality, this is typically not the case for a WLL operator. Further, the WLL operator may be less spectrum constrained than the mobile operator. The net result is that WLL operators tend to use coders with better voice quality than those used by cellular operators, by far the majority opting for the ADPCM coder.

It is certain that coders will improve relatively quickly in the coming years. A GSM-enhanced full rate coder, operating at 13 Kbits/s but with enhanced speech quality, will be introduced, and other improvements can be expected. Hence, it is likely that the data rate required for WLL speech will fall to perhaps 16 Kbits/s in three years and probably as low as 8 Kbits/s in five to ten years. Further advances beyond this would certainly be possible.

9.3   Error correction coding and interleaving

Sometimes, when you are using a mobile phone, you will notice that the speech quality "breaks up" or disappears completely for short periods of time. By moving toward a window you can sometimes improve the situation. This loss of speech quality is caused by errors. That is, the

Table 9.1
Some Voice Coders

Coder	Data rate	Quality	Comments
PCM	64 Kbits/s	Excellent	
ADPCM	32 Kbits/s	Virtually PCM	WLL applications
GSM full rate	13 Kbits/s	Noticeably worse than PCM	
GSM half rate	6.5 Kbits/s	Slightly worse than GSM full rate	Quality typically insufficient
Latest digital radio systems	4.4 Kbits/s	To be demonstrated	

transmitter might send 1011, but because of propagation problems, such as fast fading, the receiver might think that 1001 was sent. The third bit is said to be in error. This is a little like spelling something over the phone. You might say "S" but the person at the other end might respond "was that F?" An error was made because the line was not of sufficient quality. Mobile phones contain advanced systems for correcting errors that are explained here. However, as will be seen, these systems are not always able to remove all the errors. Without error correction, the speech quality would always be so terrible that you would never be able to understand the other person.

Information transmitted via a radio channel is liable to be corrupted. Interference, fading, and random noise cause errors to be received, the level of which will depend on the severity of the interference. The presence of errors can cause problems. For speech coders such as ADPCM, if the *bit error rate* (BER) rises above 10^{-3} (that is, 1 bit in every 1000 is in error, or the error rate is 0.1%) then the speech quality becomes unacceptable. For near-perfect voice quality, error rates of the order of 10^{-6} are required. For data transfers, users expect much better error rates, for example on computer files, error rates higher than 10^{-9} are normally unacceptable.

If the only source of error on the channel was random noise, then it would be possible, and generally efficient, to simply ensure that the received signal power was sufficient to achieve the required error performance without any need for error correction. However, where fast fading is present, fades can be momentarily as deep as 40 dB. To increase the received power by 40 dB to overcome such fades would be highly inefficient, resulting in a significantly reduced range and increased interference to other cells. Instead, error correction coding accepts that bits will be received in error during fades but attempts to correct these using extra bits ("redundant" bits) added to the signal.

Error correction is widely deployed in mobile radio, where fast fading is almost universally present. Error correction systems all work by adding redundancy to the transmitted signal. The receiver checks that the redundant information is as it would have expected and, if not, can make error correction decisions. An extremely simple error correction scheme would repeat the data three times. The first bit in each of the three repetitions is compared and, if there is any difference, the value that is present in two of the three repetitions is assumed to be correct. This is repeated for all bits. Such a system could correct one error in every three bits but triples the bandwidth required. Considerably more efficient schemes than this are available.

Similarly to error correction systems, there are schemes that detect errors but do not correct them. In the preceding simple example, if the message was only repeated twice, then if the repetition of a given bit was not the same as the original transmission it is clear that an error occurred but it is not possible to say which transmission was in error. In an error detection scheme, the receiver then requests that the block that was detected to be in error is retransmitted. Such schemes are called *automatic request repeat* (ARQ). They have the advantage of often reducing the transmission requirements (even accounting for the bandwidth needed for retransmission of errored blocks) but add a variable delay to the transmission while blocks are repeated. This variable delay is unsuitable for speech but typically acceptable on computer file transfer. Some of the more advanced coding systems can perform error correction and also detect if there were too many errors for it to be possible to correct them all and hence request retransmission in this case.

Error correction methods broadly fall into two types: block or convolutional coding. Both are highly involved and mathematical, and the treatment here will no more than scratch the surface. Block coding basically works by putting the information to be transmitted in a matrix and multiplying this by another matrix, whose contents are fixed for the particular coding scheme and known to both the transmitter and the receiver, as shown in Figure 9.3.

The result of the matrix multiplication forms the codeword. This codeword is then transmitted after the information, which is left unchanged. At the receiver, the information is loaded into another identical matrix, multiplied by the known matrix and the results compared with the received codeword. If there are differences, then complex matrix operations (which are processor intensive) can be used to determine where the error lies and it can be corrected. If no solution can be found, then more errors than can be corrected have occurred.

Depending on the type of matrix, codes fall into a number of families. Two well-known families are the *Bose-Chaudhuri-Hocquenghem* (BCH) family and the *Reed-Solomon* (RS) family, named after their inventors. The

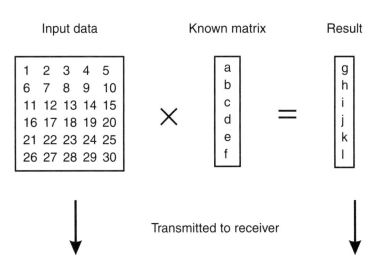

Input data   Known matrix   Result

Transmitted to receiver

Figure 9.3 The matrix multiplication used for block coding.

coding power is represented by a shorthand such as BCH(63,45,3). Such a code takes a block of 45 bits, adds 28 coding bits, and then transmits the 63-bit signal. It can correct up to three errors in the received signal, that is, an error rate of $3/63 = 4.7\%$, but increases the datastream to be transmitted by $63/45 = 1.4$ or 40%.

Convolutional coders are completely different. A diagram of a highly simplified convolutional coder is provided in Figure 9.4.

By combining the input and previously input bits in a variety of ways, predictable redundancy is added to the signal. For example, in the diagram shown, and where the + symbol represents an XOR operation, if the previous three bits had been 010 and then a new bit arrived, then the three bits in the encoder would be either 101 or 001 depending on whether the new bit was a 1 or a 0, respectively, and the only possible outputs would be 01 for bits 1 and 2, respectively, or 11 (hence if 00 or 10 was received there would have been an error). The decoder in the receiver can use this knowledge of the redundancy to correct errors. It does this using an architecture known as a Viterbi decoder. Such decoders are widely used in mobile radio. They are, however, complex to describe and beyond the scope here.

Because the coded, rather than the original, data is transmitted, convolutional codes cannot perform error detection. Therefore, they cannot

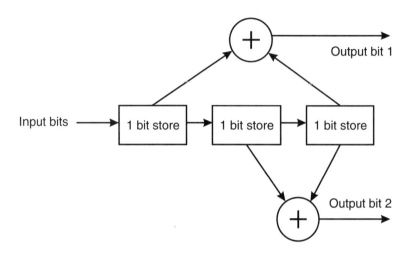

Figure 9.4   Diagram of convolutional encoder.

deploy ARQ. However, compared to block codes, they can correct more errors yet require less computing power. Often radio systems use both types of error correction systems in a concatenated arrangement in order to get the best of both worlds.

9.4   Interleaving

All error correction systems work best when the errors are randomly distributed and worst when the errors arrive in blocks. Even human error correction systems work like this. If every fifth letter of the rest of this sentence is replaced with an "error," then dou cae probwbly swill wbrk oup what hhis sags[1]. If all the errors occur in one place, then udkfgstffedng becomes more difficult. (The word in error should have been "understanding.") In order to make the speech as clear as possible, there needs to be as few errors as possible, which means that some way must be found to make the errors randomly distributed. This can be seen with the BCH(63,45,3) example. Ideally, in each block of 63 bits there would be three or less errors. The system will not work if there are zero errors in the first 3 blocks and then 12 errors in the next block.

---

1. Corrected it reads "you can probably still work out what this says."

Unfortunately, fading tends to result in errors occurring in blocks. The solution is to randomize the errors using a device known as an interleaver, which is a quite simple device. A typical interleaver would place the input bits in a matrix, filling it from left to right and then moving down a row. When the matrix is full, the data is read out in columns and transmitted. At the receiver the data fills up a matrix from top to bottom in columns and when full reads it out in rows from left to right. This is shown diagrammatically in Figure 9.5.

A block of errors would affect one column, but only one error would appear on each row, thus reducing the impact of fades.

However, interleavers have a key disadvantage; that is, they introduce delay while the matrices at the transmitter and the receiver are filled. This delay is undesirable for speech. The system designer needs to balance the reduction in error rate (and hence improvement in speech quality) with the undesirable effect of the delay.

9.5 Modulation

In Chapter 3, it was mentioned that to transmit a radio signal, it was necessary to send a changing waveform and that a sine wave was typically used for this. It was also mentioned that each user needed to transmit on a different frequency. Some way must be found for each user to encode their speech information onto their sine wave. (Otherwise, if they just send a sine wave the receiver will not be able to extract any useful

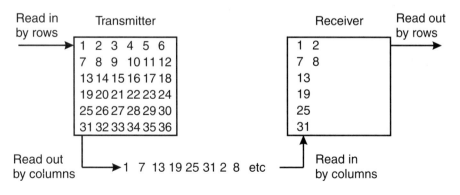

Figure 9.5 Interleaving.

information from it.) The process of encoding information onto a sine wave is called *modulation*.

There are three different classes of modulation: *frequency modulation*, *phase modulation* (PM), and *amplitude modulation* (AM)—phase modulation is strictly a class of frequency modulation and is not discussed further. Broadly, frequency modulation changes the frequency of the transmitted signal in accordance with the information to be transmitted, and amplitude modulation changes the amplitude of the transmitted signal. Sample waveforms for these two types of modulation are shown in Figure 9.6. Typically, frequency modulation is less susceptible to interference and is more widely used.

9.6 Frequency hopping

Not included in the block diagram of Figure 9.1, mostly because it has no logical place to fit there, is a technique known as frequency hopping. In frequency hopping a mobile moves from frequency to frequency while it is transmitting. As will be seen in GSM in Chapter 11, a mobile actually transmits in short bursts, each 0.5-ms long, with a gap of 3.5 ms between bursts. The reason for this behavior will become more apparent in the next section. Each burst uses a different frequency from a set of say, around 20, going through the set one-by-one and returning back to the original. Other mobiles in the cell are also hopping from frequency to frequency and the sequence that each uses is carefully designed so that all the frequencies in the cell are always in use but no two mobiles ever use the same frequency at the same time.

Frequency hopping brings two key advantages: it helps overcome fading and it reduces interference. Each of these is now explained in detail.

9.6.1 Overcoming fading

Chapter 8 discussed how there were various points in space where reflections added together such that all the signals canceled each other out. Each of these points is called a fade. The worst case would be for a mobile to stop in such a position that it was exactly in a fade. Then the signal would be all but lost and the call would be dropped. However, a fade at

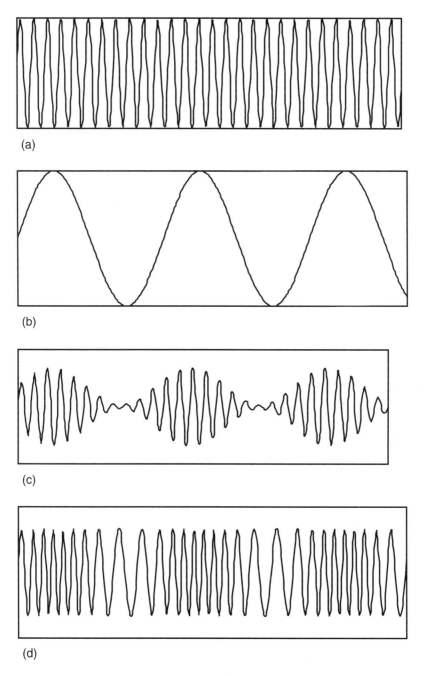

(a)

(b)

(c)

(d)

Figure 9.6 (a) Carrier waveform, (b) modulating waveform,
(c) amplitude modulated waveform, and (d) frequency modulated

one frequency is unlikely to be a fade at a different frequency because the wavelength is slightly different and hence the reflected waves are unlikely to add together so that they exactly cancel out in that particular position (although they will form a fade in some other position).

If the mobile hops from frequency to frequency, then it may be that at one frequency it is in a fade and loses all the signal. However, this will only last for 4 ms by which time it will be receiving on a frequency that is probably not in a fade. Typically, only 1 in the 20 frequencies will be in a fade at one point and hence only 1 in 20 of the bursts will be in error. Although this might result in a slight degradation in the speech, at least the call would not be dropped as it might have been without frequency hopping.

9.6.2   Reducing interference

Reducing interference is slightly more complicated to explain. Remember from Section 3.3 that the same frequencies are reused in cells far enough away so that interference will not be experienced. Now look at Figure 9.7, which shows two mobiles using the same frequencies in cells that are some distance apart. In the figure, in the best case, the mobiles are much further apart than in the worst case and hence will suffer less interference.

When designing a system without frequency hopping, the designer must assume that the worst case will occur. Hence, the designer must space cells using the same frequencies further apart than if the average case was assumed. This results in an increased cluster size (because extra cells using different frequencies are now required between the two cells using the same frequency) and hence a reduction in the overall system capacity. This is wasteful because, in most cases, the worst case will not happen. However, if it did and the cells were closer together, then unacceptable interference would result.

Now imagine that there are five mobiles in each of the cells in Figure 9.7. Each mobile hops from frequency to frequency but the sequence of frequencies used in one cell is quite different from the other. So, for example, if there were only five frequencies, then in cell 1, a mobile might use frequency 1 for the first hop; 2 for the second; then 3, 4, and 5. In the other cell a mobile might use 1 for the first hop, then 5, then

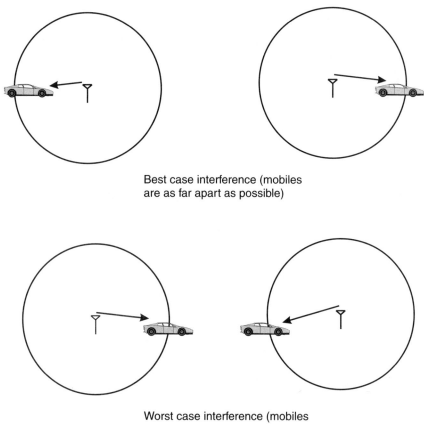

Best case interference (mobiles
are as far apart as possible)

Worst case interference (mobiles
are as close together as possible)

Figure 9.7 Best and worst interference cases.

2, then 4, and then 3. In this case, the two mobiles would only be on the same frequency for one in five hops. On the other four hops they would be on the same frequencies as other mobiles in the cells.

With five mobiles spread around each cell, the chances are that some will be in the worst case position and some in the best case position. However, for the one in the worst case position, they will only interfere with one in five hops, not all the time. This means that the interference experienced is much lower than for the worst case without frequency hopping. In practice, around 60 frequencies are used, but five was selected here to simplify the example.

Now that the designer can design for the average rather than the worst case, cells using the same frequency can be closer together, the cluster size can be smaller, and the overall system capacity can be greater. System capacity is extremely important and the additional complexity associated with frequency hopping is a small price to pay for the improvement in capacity.

9.7　Equalization

Equalization is the process of removing ISI. Essentially, in the case where there is ISI on the channel, what has happened in the channel has added together the data stream, plus a delayed version of the data stream. By taking the received signal, delaying part of it, and subtracting it from itself, the original signal can be recovered again, at least in principle.

There are many different type of equalizers that work in many different ways and to explain them all would take a book in its own right. However, understanding something about the simplest of them, the decision feedback equalizer, helps to understand some of the design parameters in radio systems.

Before it can start decoding the received signal, the decision feedback equalizer (and indeed all types of equalizer) needs to know what reflections there were and what the signal strength was from each of the reflections. In outline, this knowledge is gained by the network periodically sending a "blip" out to the mobile. Imagine that the network sent nothing for a short period, following by a short "blip," and then nothing for a bit. The mobile would then receive a series of blips, one corresponding to each reflected path arriving at the mobile and the relative strength of the blip indicating the relative strength of each reflected path. This is exactly the information it needs to set the equalizer.

Each time the mobile moves just slightly, the reflected paths change as the mobile moves through fades or as reflections appear from different angles. Hence, these blips must be sent quite frequently so that the equalizer parameters can always be set correctly. In GSM a blip is sent to each mobile once every 4 ms, or 250 times a second.

Actually, a sharp spike or blip is never sent by a radio system. Such a blip would cause tremendous interference to neighboring cells.

Remember that it was stated earlier that a sudden change in a transmitted signal requires a lot of radio frequency. Well, a spike or blip is about the most sudden change in transmitted signal possible. Instead, a special sequence of 1s and 0s is sent, called a channel sounding sequence.

A sounding sequence is a sequence of 1s and 0s, such as 011010110101. When this sequence is multiplied by a copy of itself the result is a large number (12 in this case). However, if the copy is slightly shifted compared to the original, for example, the copy becomes 110101101010, then the result of the multiplication becomes very small (two in this case).

Now you can start to understand what happens with this sequence. The transmitter sends it once to the mobile. The mobile receives a signal that is the combined result of all the reflections. Every bit period the receiver multiples what it has received by its copy of the sounding sequence. Each time it aligns with a reflected signal the receiver will get a large result, or a "spike" from the multiplication. By looking for when these spikes occur and their size it can deduce the delay and strength of the reflected paths.

A simplified diagram of a decision feedback equalizer is shown in Figure 9.8, where the terms used in the diagram are explained in the remainder of this section. Here is how it works in outline. Imagine that

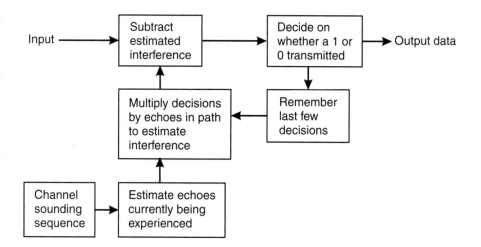

Figure 9.8 Simplified decision feedback equalizer.

the mobile receives two signals. One comes directly from the transmitter and the second is reflected off a distant building so that it arrives two bits later and with only half the signal strength.

Now imagine that nothing has been transmitted for a while. When the first bit is transmitted, it is received directly by the mobile and can be decoded. The mobile remembers it. When the second bit is received the same thing happens. When the third bit is received the mobile knows that it will actually be the sum of the third bit (from the direct path) and the first bit, at half power (from the reflected path). So first, the mobile subtracts a half-strength version of the first bit from what it has received. It can then decode the third bit and remember it. When the fourth bit arrives, it subtracts a half-strength copy of the second bit from it and decodes the bit and then repeats this process through the end of the transmission.

This works fine as long as no error is made in decoding a bit. Say, however, the third bit was actually decoded in error because of a sudden burst of noise. When the fifth bit arrives and the third bit is subtracted from it, the wrong information is actually subtracted from the fifth bit. This means that the result will be quite different from what was transmitted and the mobile will probably decode the fifth bit in error. And so it goes on, with the error propagating throughout the received signal until such time as it can be corrected. Correction only occurs when the transmitter stops transmitting for a short period, the echoes die away, and the equalizer can sort itself out again. This is the reason why the decision feedback equalizer is rarely used. In GSM an equalizer called Viterbi equalizer is used that adopts the same sort of principles but is too complex to describe here.

9.8   Multiple access

Remember that each operator has a certain number of radio frequencies that they need to divide amongst their users whenever the user wants to make a call. The most intuitive way to do this is to divide the total radio frequency into a number of small frequency bands and to give a user one of these bands whenever they want to make a call. Indeed, this is exactly what was done for most of the history of mobile radio. This approach is called *frequency division multiple access*. The "multiple access" part derives from allowing multiple subscribers to have access to the radio spectrum. The "frequency division" part says that the spectrum is divided into lots of pieces with different frequencies.

Frequency division multiple access was the key multiple access technique until around 10 years ago. Then, for reasons that are subsequently explained, other ways of dividing the spectrum were explored. A second approach is to say to each user that they can have all the radio spectrum but only for, say, one-tenth of each second. They transmit from 0 to 0.1 of a second, from 1 to 1.1, from 2 to 2.1, and so on. In that time they must transmit a whole second's worth of information, so they need to send information faster. This requires more spectrum, but that works out because they have access to a lot more spectrum. This approach is called time division multiple access because the spectrum is divided in time, with different users accessing all the spectrum but at different times.

Finally, and most recently, a technique for allowing everyone who wanted to transmit to use all the spectrum all at the same time was invented. This sounds crazy, because you would expect them to interfere terribly with each other. However, because each user has much more spectrum than they need (because they now have access to all the spectrum rather than just a small part), they can add a lot of additional information to the voice information, which allows the receiver to extract their signal from the interference. This additional information is called a *code*, and hence the technique is known as code division multiple access.

So, in summary, there are broadly three ways to divide up the resource (the radio spectrum) amongst those wanting to make a call:

- *Frequency division multiple access* (FDMA), where the frequency band is divided into a number of slots and each user accesses a particular slot for the length of their call;

- *Time division multiple access* (TDMA), where each user accesses all the frequency band but only for a short period of time;

- *Code division multiple access* (CDMA), where each user accesses all the frequency band for all the time but distinguishes their transmission through the use of a particular code.

There are cellular technologies available that make use of each of these different access methods. Their advantages and disadvantages are now described in more detail.

9.8.1 FDMA

In a typical FDMA system, the available bandwidth would be divided into slots about 25-kHz wide as shown in Figure 9.9. Each slot contains a dedicated transmission. The problem lies in the fact that the transmitted power when plotted against bandwidth is not an idealized rectangle (i.e., equal power is not transmitted across the whole available bandwidth). Figure 9.10 shows the spectrum transmitted by the GSM mobile radio system, which uses one of the most complex filtering arrangements of any technology available today. Two adjacent channels are shown, and it is clear that there is significant interference between the two channels.

It is clear, when compared to an ideal rectangular spectrum emission, that the GSM system does not transmit as much power within the band, the power levels falling away as the band edges are reached. This represents an inefficiency, reducing the power available to the mobile from the ideal case.

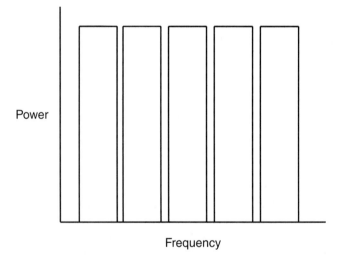

Figure 9.9   An ideal FDMA spectrum.

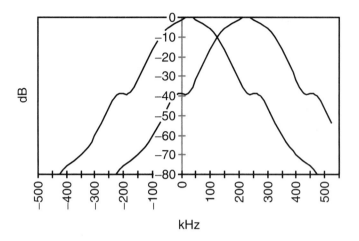

Figure 9.10   GSM spectrum.

A more pragmatic problem with FDMA is that at the base station, each individual channel requires a separate power amplifier before passing through an expensive high-power combiner and then being transmitted from the antenna. It would be possible to combine the signals before the amplifier if the amplifier was highly linear, but in practice such amplifiers are extremely expensive and inefficient in their use of power.

In summary, the advantage of FDMA is that it is the simplest access method to implement; while the disadvantages are the loss of efficiency caused by imperfect filtering and expensive *radio frequency* (RF) elements required at the base station.

Now that technology has advanced to the point where other access methods can be implemented at relatively low cost, the advantage of FDMA is broadly negated. For this reason, virtually no digital systems use FDMA, this access method is only in use in the older analog technologies.

9.8.2   TDMA

In TDMA a user has access to a wide bandwidth but only for a short period of time. Again, using the example of GSM, which is a TDMA system, a user has access to 200 kHz of bandwidth for one-eighth of the time. To be more precise, the user has access to the channel for 577 μs every 4.6 ms.

During that period they send a burst of data that was previously buffered in the transmitter.

Like FDMA, TDMA has its inefficiencies that are caused by the need to allow the mobile time to increase its power from zero and reduce it back to zero again (this is called *ramping up* and *ramping down*). If time is not allowed for this transition, then the near-instantaneous change in power results in a momentary use of an extremely high bandwidth, with resulting interference to a wide range of users. Guard bands are provided in order to allow for this ramping up and down. What happens when a user transmits in a TDMA system is shown in Figure 9.11, where it can be seen that approximately 30 μs is required for the user to ramp up to the full transmitted power for a burst of 540 μs. (The time taken to ramp down is also used by the next mobile to ramp up so it is not required in addition to the ramp up time.) Hence, the inefficiency is 30/540 = 5.5%.

TDMA systems require additional overheads because they have to send timing information so that the subscriber units know exactly when to transmit. Another problem is that by transmitting at a higher data rate (even though only for a shorter time), the problem of ISI is exacerbated. In some systems, such as GSM, this has resulted in the need for an

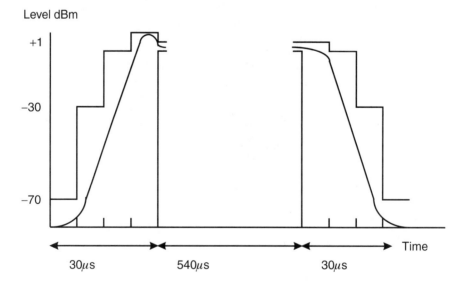

Figure 9.11   GSM ramping up for a burst transmission.

equalizer as explained in Section 9.7, that removes the ISI in the receiver. Remember that in order for an equalizer to work, the channel characteristics need to be periodically measured. This is achieved by placing a sounding sequence in the middle of each burst. However, the sending of this sounding sequence represents a substantial inefficiency (22%) in the use of radio spectrum.

TDMA overcomes one of the key disadvantages of FDMA. Because only one user's signal is transmitted at any one time, only a single amplifier is required at the base station and the need for combiners is removed. This provides a substantial cost saving in the base station. TDMA is also, generally, more spectrum efficient than FDMA, because the size of the guard band relative to the size of a burst is much smaller than the size of the guard band effectively required for FDMA relative to the bandwidth of an FDMA channel.

In summary, the advantages of TDMA are the more efficient use of the radio spectrum than FDMA and the fact that it is less expensive to implement than FDMA; while the disadvantages are that more complex subscriber units are required and ISI may become problematic.

TDMA is a widely used multiple access method for many mobile systems available to date including GSM. As an aside, strictly most TDMA systems are actually TDMA/FDMA. For example, GSM places eight users on a 200-kHz channel using TDMA but then divides the assignment into 200-kHz slots using FDMA.

9.8.3 CDMA

This section goes into somewhat more detail on CDMA because it is more complex and less intuitive than the other access technologies. Further, as will be seen, there is significant debate as to whether CDMA or TDMA is more efficient of radio spectrum and an understanding of this debate requires a good understanding of CDMA itself.

CDMA transmits on all the frequencies for all of the time. The result of this is that all the users interfere. However, they each use a dedicated code that they have been given by the transmitter to help them pick out their signal from the interference.

In CDMA, the user first generates their data, which could be, for example, the output of a speech coder. This is generated at a rate known as the bit rate, which as was seen earlier, might typically be around 16 Kbits/s. Each bit is then multiplied by the unique code that has been assigned to that particular mobile by the network and the result is then transmitted. A typical codeword might be 64-bits long and the transmitter would send this codeword whenever the bit from the speech coder was a 1 and would send the inverse (i.e., 0s replaced by 1s and 1s replaced by 0s) whenever the bit from the speech coder was a 0. Now 64 bits have to be sent whenever 1 bit is generated by the speech coder, so the data rate becomes 16 Kbits/s × 64 = 1,024 Kbits/s. The mobile uses 64 times as much spectrum as it would have done for FDMA. This higher rate of 1,024 Kbits/s is known as the *chip rate*. The process of multiplication by a codeword is known as *spreading* and is shown for an example datastream and codeword in Figure 9.12. The process of generating the signal is shown in Figure 9.13, where the circles with crosses in the middle mean "multiply."

At the receiver a process called *despreading* is required to recover the data. This involves the multiplication of the received signal with the same codeword. Such multiplication results in the original binary information being decoded but with an enhancement of the signal level by a factor equal to the length of the code (e.g., 64 in the example used). This enhancement allows 64 times as much interference to be tolerated on the link as would be the case if the voice information had not been multiplied by the code. It is by this means that the interference from all the other users transmitting on the same frequency can be overcome.

Each other user has their own code for the length of time that they make the call. In the same way that with FDMA each user needs a

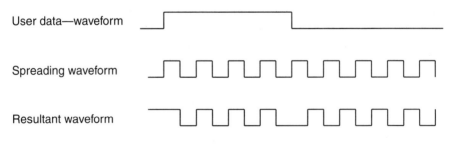

Figure 9.12   Generation of a CDMA signal.

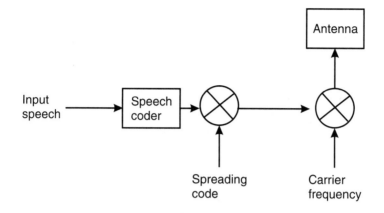

Transmitter—the spreading or coding process

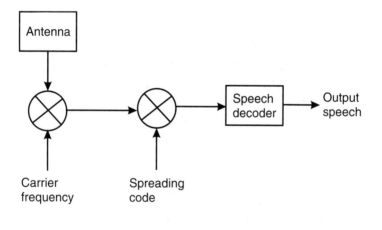

Receiver—the de-spreading process

Figure 9.13 Example of the generation of a CDMA signal.

different frequency, with CDMA each user needs a different code. If two users had the same code, then the receiver would not be able to differentiate between them and the interference would be severe. In order to get the best performance, it is necessary to select special codes, as discussed in the next section. However, there is a limited number of these special codes; in fact, there are as many special codes as there are bits in the codeword, for example, 64 in the previous example.

This shortage of codewords can be a problem in some systems, particularly cellular, where interference from adjacent cells can be expected. The way this is overcome is part of the next section.

The advantages and disadvantages of each of the access methods is summarized in Table 9.2.

▼   The use of spreading codes in CDMA

In an isolated CDMA cell (i.e., one without any neighbors), each user's signal is spread by a particular spreading code. Spreading codes should be *orthogonal* so that they do not generate interference. Two codes are said to be *orthogonal* to each other if, when multiplied together over the length of the spreading code, and the total summed, the result is zero, regardless of whether the code was carrying user data corresponding to a 1 or a 0. This is best illustrated by a sequence of examples. In order to make these fit on the page, a spreading factor of 8 has been used; that is, for every bit of user information, eight bits of the spreading code are transmitted. In practice, a spreading factor of typically around 64 would be used. In the

Table 9.2
Summary of the Different Multiple Access Methods

	Advantages	Disadvantages	Systems using the access method
FDMA	Simplicity.	Inefficient of spectrum. Expensive RF elements required.	Analog cellular systems.
TDMA	More efficient than FDMA. Less expensive than FDMA.	Mobiles are more complex. ISI becomes problematic.	Most digital cellular systems including GSM and D-AMPS.
CDMA	Highest spectrum efficiency.	Highest complexity.	Only CDMAone - described later.

examples, the spreading waveforms have been shown as sequences of 1s and −1s while the user data has been shown as 0s and 1s. This is shown diagrammatically in Figure 9.14.

Moving to this numerical representation it is now possible to examine how orthogonal codes work in more detail. First, consider how the receiver decodes the speech data by multiplying the received signal by a copy of the spreading sequence held in the receiver, and assume that the spreading sequence is 1,−1,1,−1,1,−1,1,−1 as shown in Figure 9.14. To understand the process, look at the first column in Table 9.3. The transmitted data (from the speech coder) is a 1, which means that the spreading sequence is transmitted directly; that is, the top entry in the first column is the first bit of the spreading sequence, namely 1. The receiver codeword is the same spreading sequence and the first bit of this is, as before, a 1. When multiplied together the result is a 1. When all the columns are added together the result is 8.

When a 0 is transmitted, the inverse of the spreading sequence is sent. This means the first bit (as shown in the first column, fourth line) is a −1.

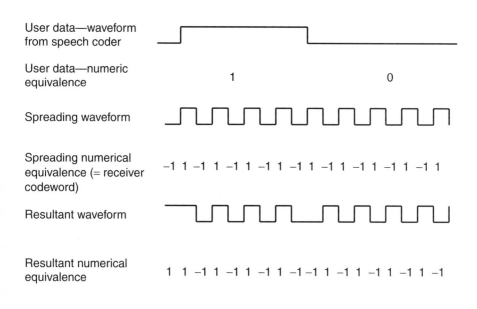

Figure 9.14 Numeric representation of CDMA waveforms.

Table 9.3
CDMA Transmission Without Other Users

Transmit (data = 1)	1	−1	1	−1	1	−1	1	−1	
Receiver codeword	1	−1	1	−1	1	−1	1	−1	
Multiplication	1	1	1	1	1	1	1	1	= 8
Transmit (data = 0)	−1	1	−1	1	−1	1	−1	1	
Receiver codeword	1	−1	1	−1	1	−1	1	−1	
Multiplication	−1	−1	−1	−1	−1	−1	−1	−1	= −8

The receiver codeword is still a 1. The result of the multiplication is −1. When all the columns are added together the result is −8.

The effect is that when the received waveform is multiplied by the receiver codeword and the result added up over the period of one bit of user data, the output of the correlator is either 8 or −8, corresponding to a "1" or a "0" in the user datastream. (If a spreading code of length 64 had been used the output would have been either 64 or −64.)

Now it so happens that a spreading sequence of 1, 1, −1, −1, 1, 1, −1, −1 is orthogonal to the spreading sequence used in Table 9.3. Using the same multiplication process as previously, but now assuming that there are two subscribers transmitting at the same time, the multiplication shown in Table 9.4 occurs in the base station.

Here, in the first column, the transmit data is the first bit of the wanted spreading sequence, a 1. The column marked interferer gives the first bit of the spreading sequence that the other subscriber transmitting at the same time is using, also a 1. The result of $1 + 1 = 2$, which is the received signal. This is multiplied by the receiver codeword, the first bit of which is a 1, as before. The result of $2 \times 1 = 2$. In the bottom half of the table the same operation is performed but with the transmit data being a 0 and, hence, the inverse of the spreading sequence being transmitted.

It can be seen that the output from the multiplication is the same as in the first case when there was no interference. Next consider the situation where the interferer is sending user data corresponding to a 0, as shown in Table 9.5.

Table 9.4
CDMA Transmission With a Second User Sending a "1"

Transmit (data = 1)	1	−1	1	−1	1	−1	1	−1	
Interferer (data = 1)	1	1	−1	−1	1	1	−1	−1	
Received signal	2	0	0	−2	2	0	0	−2	
Receiver codeword	1	−1	1	−1	1	−1	1	−1	
Multiplication	2	0	0	2	2	0	0	2	= 8
Transmit (data = 0)	−1	1	−1	1	−1	1	−1	1	
Interferer (data = 1)	1	1	−1	−1	1	1	−1	−1	
Received signal	0	2	−2	0	0	2	−2	0	
Receiver codeword	1	−1	1	−1	1	−1	1	−1	

Table 9.5
CDMA transmission With a Second User Sending a "0"

Transmit (data = 1)	1	−1	1	−1	1	−1	1	−1	
Interferer (data = 0)	−1	−1	1	1	−1	−1	1	1	
Received signal	0	−2	2	0	0	−2	2	0	
Receiver codeword	1	−1	1	−1	1	−1	1	−1	
Multiplication	0	2	2	0	0	2	2	0	= 8
Transmit (data = 0)	−1	1	−1	1	−1	1	−1	1	
Interferer (data = 0)	−1	−1	1	1	−1	−1	1	1	
Received signal	−2	0	0	2	−2	0	0	2	
Receiver codeword	1	−1	1	−1	1	−1	1	−1	
Multiplication	−2	0	0	−2	−2	0	0	−2	= −8

So whatever is transmitted by the interferer and by the wanted user, the receiver produces the same result as if there were no interferer. Clearly the signals are orthogonal. More orthogonal interferers can be added without having any effect on the wanted signal. Even if the interferers are received with different power from the transmitter, the correct result is still achieved as demonstrated in Table 9.6.

Table 9.6
CDMA Transmission With a Second User Transmitting at a Higher Power

Transmit (data = 1)	1	−1	1	−1	1	−1	1	−1	
Interferer (data = 1)	2	2	−2	−2	2	2	−2	−2	
Received signal	3	1	−1	−3	3	1	−1	−3	
Receiver codeword	1	−1	1	−1	1	−1	1	−1	
Multiplication	3	−1	−1	3	3	−1	−1	3	= 8
Transmit (data = 0)	−1	1	−1	1	−1	1	−1	1	
Interferer (data = 1)	2	2	−2	−2	2	2	−2	−2	
Received signal	1	3	−3	−1	1	3	−3	−1	
Receiver codeword	1	−1	1	−1	1	−1	1	−1	
Multiplication	1	−3	−3	1	1	−3	−3	1	= −8

The only time when this relationship does not hold is if the received signals are not synchronized so that the changes from 1 to −1 or −1 to 1 in the spreading sequences occur at slightly different times for the two sequences. This would occur when the subscriber units are different distances from the base station and so experience different propagation delays, although there are techniques that can be used to help this, as discussed in Chapter 11.

The codes that were used previously are actually two of the so-called Walsh code family, which is the most widely used sets of CDMA codes in cellular applications. The complete family of codes is given in Table 9.7.

All of these codes are orthogonal to each other. This can be seen by multiplying any of the codes by any other code and summing the total; the result will always be 0.

In a single cell in isolation, each channel can be given a separate Walsh code and the maximum system capacity can be reached. The problem comes when there is a number of neighboring cells using the same frequency. In order to prevent excessive interference, all the users in all the neighboring cells need different spreading sequences. However, there are only as many sequences as the code length, whereas there will be many more interfering users than this. Of course, one solution would

Table 9.7

The Family of Walsh Codes of Length 8

Code 0	1	1	1	1	1	1	1	1
Code 1	1	1	1	1	−1	−1	−1	−1
Code 2	1	1	−1	−1	−1	−1	1	1
Code 3	−1	−1	1	1	−1	−1	1	1
Code 4	1	−1	−1	1	1	−1	−1	1
Code 5	1	−1	−1	1	−1	1	1	−1
Code 6	1	−1	1	−1	1	−1	1	−1
Code 7	−1	1	−1	1	1	−1	1	−1

be not to let neighboring cells use the same frequency, but the result of this would be a much lower system capacity than could otherwise be the case. For a long time, this problem was thought by researchers to mean that CDMA could not be used for commercial applications.

The breakthrough came with the realization that if nearly orthogonal codes were used rather than fully orthogonal codes that the interference between different users would not be too severe, although greater than for the orthogonal case. There is a family of near-orthogonal codes called pseudo-noise, or PN codes. The good thing about PN codes is that there are literally millions of them, overcoming the limitations with Walsh codes.

A PN code is a sequence of 1s and 0s that repeats periodically. PN code sequences have the property that if multiplied by themselves the result has the same magnitude as the length of the sequence (in the same way that the orthogonal codes of length 8 resulted in a multiplication of +/− 8). If multiplied by the same sequence but shifted in time by any number of bits, the result is −1 (unlike orthogonal codes where the result is 0), which means that some interference is generated.

In the mobile systems there is a tendency to give each user a separate PN code since they could roam into any cell and there is a need to ensure that they do not generate interference wherever they are.

▲

9.8.4 CDMA power control

One of the main concerns with CDMA is power control. Mobile radios are not very good at transmitting just the right amount of power. To do this accurately requires complicated and expensive circuits, whereas mobile phones are designed to be relatively inexpensive devices. The net result is that mobiles cannot be relied upon to transmit with exactly the power that they have been told to. For most mobiles, transmitting somewhere between half as much and twice as much power (plus or minus 3 dB in engineering parlance) as required is quite a challenge!

If a system using nearly orthogonal codes could normally accommodate eight users and one user transmits at four times the level than he needs to, then only three other users can be tolerated, that is, the capacity of the system has been halved. In a cellular environment, some of the subscriber terminals will be closer to the base station than others. In order to maximize the system capacity it is important that those closer to the BS transmit with a lower power so that all signal levels are received with the same signal strength.

The accuracy required for this power control is very high. A 3-dB error (i.e., a unit working at twice the intended power) would halve the capacity of the system. In cellular CDMA systems, such accuracy is very difficult to achieve as mobiles pass through a fast fading environment. The dramatic effects of errors in the average power transmitted by all the subscriber units is shown in Figure 9.15. This figure shows that if all the mobiles had an average error of 3 dB, the system capacity would only be around 40% of that predicted. Achieving this accuracy is actually quite difficult.

Within the mobile community, there has been much impassioned discussion about whether TDMA or CDMA represents the best access technique. Because of the importance of this debate, the entire next chapter has been set aside to address it.

9.9 Packet and circuit switching

The types of systems discussed to date have implicitly assumed circuit switching. That is, when a user starts to make a call, a circuit is established between the user and the network, which is maintained for the duration

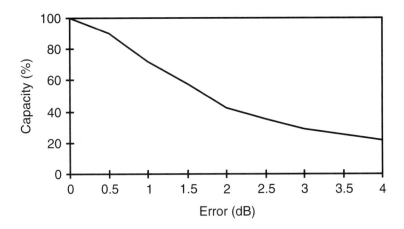

Figure 9.15 Sensitivity to power control error in CDMA systems.

of the call. This circuit may be an FDMA channel, a timeslot on a TDMA channel, or a CDMA orthogonal code; the net effect is that nobody else is able to use that particular resource for the duration of the call.

An alternative to circuit switching is something called *packet switching*. In a packet-switched system, a user is not given their own radio channel for the duration of their call. Instead, the subscriber unit collects data from the user until its buffer is full and then requests a channel from the network for a short period of time, typically just a few milliseconds, to transmit the packet of data. The mobile then relinquishes the network resources and waits for the buffer to fill again. This is beneficial because while the user is not speaking the spectrum resources can be used by other users, but it adds significant complexity to the system and can result in delays that are annoying for speech transmission.

Packet switching comes in two forms: connection oriented and non-connection oriented. In the case of connection oriented, something termed a *virtual circuit* is established between the transmitter and the receiver. A virtual circuit is an agreed-upon path that packets of data will follow from the transmitter to the receiver whenever they are sent. However, no spectrum resource is set aside just for this, each packet needs to ask for resource. Because there is no dedicated circuit, this is called a virtual circuit. This virtual circuit is set up when the first packet is received. All subsequent packets received for the same destination travel via the same route. Because they all follow the same route, they will be received in the order in which they are transmitted.

In the case of nonconnection-oriented packet switching, each packet of data is treated as if no previous packet had been sent. Potentially, a packet could be sent via a different route from the previous packet. This might occur if one of the switches became congested and asked that packets temporarily be routed via a different switch. Because of this the packets might not arrive at the receiver in the order that they were sent. The receiver then has to store the incoming packets until it has received those sent earlier than the last one received. It can then correctly order the data prior to presenting it to the user.

In outline, circuit switching provides a low and known delay but uses resources inefficiently compared to packet switching. Broadly, circuit switching is suitable for voice while packet switching is suitable for data. Packet switching is unsuitable for voice because the delays suffered by each packet can be variable, resulting in significant and unwanted voice delay. Packet switching is particularly suitable when the data to be transmitted arrives in short bursts and when short delays can be tolerated. This is illustrated by an example.

Imagine a data source that provided data at the rate of 200 bits every 3 sec and required that the delay on transmission was less than 2 sec (e.g., vehicle location systems provide data in this manner). If this was transmitted via a circuit-switched channel, with, say, a data capacity of 9.6 Kbits/s, a call set-up time of 1.5 sec, and call clear-down time of 1 sec, then it would be necessary to maintain the data channel dedicated to this use. This is so because if the attempt was made to clear down the call between bursts, the signaling required to do this would take so long that the subscriber unit would need to immediately re-establish the channel once it had been cleared down. However, using a packet protocol, with an overhead of 20%, only $200/3 \times 20\%$ bits/s = 80 bits/s would be transmitted. This requires less than 1% of the available channel capacity. Packet mode systems are also ideal for asymmetrical applications when more data is transmitted in one direction than the other (e.g., Internet browsing). Because the uplink and downlink need not be paired, uplink resources are freed for another user who may wish to send a large data file in the uplink direction but receive little in the downlink direction.

The ideal radio system would probably include both circuit- and packet-switched capabilities. Indeed, some modern mobile radio systems, including GSM, are being developed with such dual capabilities.

Problem 9.1

What are the advantages and disadvantages of low bit rate speech coders?

Problem 9.2

Why are interleavers required and what do they do?

Problem 9.3

What are the two advantages of frequency hopping and explain simply why they occur?

Problem 9.4

In CDMA, why does the transmitted sequence contain more bits than the data the user wants to send? By what factor is this transmitted sequence larger than the data?

Problem 9.5

What are orthogonal CDMA sequences and why are they desirable?

10

TDMA or CDMA

10.1 Introduction to the debate

If you have been at all involved in the world of mobile radio over the last five years or so, you will probably at some time have come across an article talking about whether CDMA was better than TDMA or whether the Qualcomm IS-95 cellular standard (now known as CDMAone) was better than the GSM cellular standard. If you read the article, you probably were not a lot wiser as a result. If you have not come across the debate between CDMA and TDMA, then you may be wondering why this subject is sufficiently interesting to merit a chapter in its own right.

Basically what happened is that what was a small American company called Qualcomm (it is quite large now), amongst whose founders are two of the most eminent engineers in the world of mobile radio, Jacobs and Viterbi, suddenly announced to the world in 1991 that it had invented a new cellular system based on CDMA and that the capacity of this system was 20 or so times greater than any other cellular system in existence. Of course, the world took note, especially given the high regard in which the inventors of this system were held. However, not all of the world was particularly pleased by this apparent breakthrough—in particular, GSM manufacturers became concerned that they would start to lose market share to this new system.

It soon became clear that it was too complicated to actually work out the capacity of this CDMA system; and although much simulation and mathematical work was performed, this tended to be inconclusive and easily attacked by opponents. Qualcomm by this time had staked their whole company on the success of this technology while the GSM manufacturers had invested millions in GSM that they wanted to see recouped through sales. The result was continual and vociferous argument between Qualcomm and the GSM manufacturers. It was all a little pointless because until CDMA systems were deployed nobody could really tell which system was best but neither side wanted to lose the propaganda war. The debate has taken on such a significance that it is worth devoting a whole chapter to explaining it in detail.

It should be noted that after six years and the application of most of the finest minds in mobile radio, this debate has still not been resolved fully. It would be somewhat optimistic to expect this chapter to provide all the answers. What it can do is put the discussion on an independent and rational footing and note the key issues and implications.

10.2   A more general look at CDMA versus TDMA

The whole debate about TDMA and CDMA tends to confuse almost all observers, even engineers, mostly because of the complexity of CDMA. But when discussed intuitively, some of the issues become much clearer. TDMA and CDMA are both ways of dividing up the radio spectrum so that a number of users can talk at the same time. You could think of them as two ways of cutting a cake, one in pie-shaped slices from the center and the other in long thin slices across the cake. Both result in the same amount of cake divided between the same users. So at a very crude level, think of them as two different ways of dividing up the spectrum. This would suggest that the selection of a particular one of CDMA or TDMA is not of great importance.

Two factors result in the discussion about the division of the cake becoming less academic. The first is the loss during the division process. Imagine that the knife used to slice the cake was dirty and, as a result, the edges of each slice had to be discarded. The relative amount that has to be discarded with the two different techniques then becomes a differentiating factor between them. The second factor is less easy to explain by analogy. Remember that cellular systems achieve their capacity by reusing the spectrum and that the ability to reuse the spectrum as efficiently as possible is a key factor in the system capacity, as expressed in the cluster size. Remember also that frequency hopping allowed an increased capacity by making sure that one mobile did not suffer a lot of interference while another one had hardly any interference. In fact, if the interference is distributed evenly across all mobiles, the maximum capacity results.

It is worth pausing to examine the last sentence in more detail because it is actually a fundamental, although little understood, design concept in mobile radio systems. Imagine cell A using 10 frequencies and cell B, some distance away, also using the same 10 frequencies. You want to be sure that the interference on each of the frequencies is sufficiently low that reliable communications can be made. Put together, the mobiles in cell A are going to generate a certain amount of radio signal. If this is not evenly divided across all frequencies, then one mobile in cell B will receive more interference than the others. In order to make sure that this mobile can communicate correctly, the cell needs to be far enough away that this interference level is acceptable. However, this implies that on the other nine frequencies, the interference level is even lower. This is great for those mobiles who get a better quality, but it means that on those nine frequencies the frequencies could have been reused closer to cell A, resulting in a lower cluster size and greater spectrum efficiency. Only if the interference is evenly distributed across all the mobiles can the minimum cluster size be adopted and the maximum capacity achieved.

So when comparing CDMA and TDMA systems, the real questions to ask are:

- How much spectrum is lost in the process of division of the frequencies?

- How successful is the access method in distributing interference evenly across all the users?

10.2.1   Division of the frequencies

In TDMA, some of the frequencies are lost due to the following reasons.

- As discussed in Section 9.8, there is a need to allow mobiles time to increase their power at the start of the burst.

- Because of the high bandwidth used, ISI becomes problematic and space needs to be set aside for a sounding sequence for the equalizer.

- Information needs to be sent to the mobiles telling them where the TDMA frames start and stop to assure that they transmit in the correct place.

Of these, the first two are most important. As will be seen in Section 11.2, in a typical burst of 148 data bits, only 116 can be used to transmit the user's data, representing an efficiency of only 78%.

In CDMA, the inefficiency is better hidden. It actually results from signals not being received with equal power. The difference in power represents an inefficient division of the spectrum amongst the different codes. As shown in Section 9.8, with a relatively small error in power control, the system capacity might fall to only 40% of what it would be with perfect power control, although this is still to be fully demonstrated in real life systems.

So as regards the division of the radio frequencies, TDMA would seem to be slightly better than CDMA.

10.2.2   Distributing the interference

CDMA inherently evenly distributes interference because all the mobiles in all the cells use the same frequencies and so all interfere equally with each other. The very nature of a CDMA signal looks just like noise, and so interference is of a form that is least harmful. Discontinuous transmission, as explained toward the end of this section, simply reduces the interference that remains evenly distributed.

TDMA inherently results in uneven interference as different mobiles use different timeslots and different frequencies and, depending on where mobiles are in the cell, some will interfere with each other more strongly than others. This concept was discussed in Section 9.6.

If nothing is done to try to spread the interference more evenly in TDMA, then CDMA achieves cluster sizes of around one-tenth of those of TDMA, dramatically overcoming the slight disadvantage caused by power control.

So overall, if no special measures are adopted in TDMA, then CDMA has a much higher capacity, perhaps between 5 and 10 times, purely as a result of the more even interference and hence smaller cluster size.

10.2.3 Making TDMA interference more even

One of the key techniques for making the interference more even was already examined in Section 9.6, namely, frequency hopping. Another technique that provides very significant gains is known as *dynamic channel allocation*.

▼ **Dynamic channel allocation**

In GSM, all the frequencies available are divided up between the different cells. So cell A might be assigned frequencies 1,11,21,31, for example. This remains fixed more or less for all time. A more efficient approach results if frequencies are not divided between the different cells. Instead, whenever a user requests a channel, the cell makes a quick measurement of the interference on all the frequencies available to the system and selects the one with the lowest interference.

This brings a number of advantages. Should there be one frequency with a particularly high interference, this can be avoided and left for the next cell to use. Should a user in another cell be generating lower interference than normal, perhaps because they are behind a building, a frequency that could not normally be used can now be employed. Should there be more users than normal in one cell, frequencies can be "borrowed" from neighboring cells to overcome the congestion. *Dynamic channel allocation* (DCA) does not make the interference more even, it accommodates uneven interference better, resulting in a cluster size that might be only half that for a fixed channel allocation strategy.

Dynamic channel allocation is used in most cordless systems, not because of the increase in capacity, but because each user owns their own base station; hence, frequency planning is not possible. In this case, the base stations must seek the best frequencies—there are no other options. It is likely that all future TDMA cellular systems will have dynamic channel allocation.

▲

This discussion hopefully made the issues in the comparison of CDMA and TDMA clearer; what it did not do is say which one is best—this is shown subsequently. It is best to think of these as simply two different ways of cutting the cake and that the decision between which one to use, while important, is not an overriding concern.

10.3　Comparison of advantages and disadvantages

10.3.1　Capacity

The main debate between CDMA and TDMA concerned which provides the greatest capacity. It turns out that it is almost impossible to calculate the capacity of CDMA systems; the mathematics is just too complicated. As experience has subsequently shown, those who did try to calculate it generally got it wrong. There are actually TDMA systems and CDMA systems working and the easiest approach is simply to compare the capacity of each in real life. It turns out that CDMA systems currently provide a capacity probably around 30% greater than TDMA systems[1]. This may change in the future as new CDMA and TDMA techniques are introduced, but it seems likely that CDMA systems will maintain a small but not insignificant capacity increase into the foreseeable future.

In addition to capacity, a number of other advantages are claimed for CDMA systems. Each of these is examined in this section to assess whether the claims can be supported.

---

1. Actually, this is a gross simplification. Deployments of the same system in different networks result in different levels of capacity depending on trade-offs made by the operators. There are only a few CDMA systems currently in existence, so it is probably too early to derive an average of the CDMA capacity. The figure given here is based on early indications from a limited number of CDMA networks.

10.3.2   Greater range

It is claimed that CDMA systems have a greater range than equivalent TDMA systems. Range is related to the minimum signal level that the receiver can reliably decode, so the claim is basically one that CDMA can work with a lower received signal strength than TDMA. The previous section showed that this is, of course, true, since the receiver applies a gain equal to the spreading sequence length to the received signal with a CDMA network but not with a TDMA network.

10.3.3   Sectorization

Sectorization was discussed in detail in Chapter 8, where it was noted that it was the division of a circular cell into a number of pie-shaped sectors. It is claimed that if this is performed in a CDMA network, the same frequency can be used in each sector, increasing the capacity of the system by the number of sectors deployed. This claim is correct. It is also claimed that using sectors in a TDMA arrangement does not increase capacity. This is also broadly correct. Fundamentally, when a cell is sectorized, the cell radius remains the same. Hence, the transmitted power remains the same and, hence, the distance required to the next cell using the same frequency also remains the same. However, there are now more cells within this sterilization radius, and hence more frequencies need to be found to avoid interference. So although the sector is smaller than the cell, and hence has to support less traffic, it also has fewer frequencies with which to do this (because the total frequency assignment has been divided by a larger cluster size). This is not the case with CDMA where using the same frequency in adjacent sectors only slightly increases the interference to neighboring cells and slightly reduces their capacity.

TDMA could achieve a real gain if, instead of making a cell into a number of smaller cells by sectorizing it, it was divided into smaller circular cells, that is, the base stations were distributed around the cell and transmitted on a lower power level. This approach results in similar equipment costs but much higher site rental and backhaul costs; hence, it tends to be avoided except where absolutely necessary.

In summary, the CDMA capacity can be increased by a factor of 2 to 3 by sectorization with only a small increase in cost. This option is not available to TDMA and, hence, represents an advantage to CDMA.

▼   Discontinuous transmission

When calculating the capacity of both TDMA and CDMA systems, one of the important factors is whether *discontinuous transmission* (DTX) is used. When you make a phone call, you do not typically speak all the time. On average, you will talk half the time and the other person half the time. Even when you are talking, there are short gaps between words. It is typically estimated that a person only talks for 40% of the time. When you are not talking, there is little need to transmit anything. By not transmitting, less radio spectrum is used and in principle the system capacity can be increased. However, achieving this increase is not always simple.

In the case of GSM, there is a circuit in the mobile known as a *voice activity detector* that detects when you are not speaking. Voice activity detection is itself a complex subject, since detectors can be easily confused by background noise or quiet speakers. When the voice activity detector decides that you are not speaking, it sends a message to the network saying that the user is not speaking and stops transmitting the user's voice. It also sends a sample of background noise. The reason for this is that otherwise, the person to whom you were talking would suddenly hear all the background noise disappear, which is very disconcerting and tends to lead people to think that the call has been dropped. The network remembers the background noise sample and repeats it until such time as you start talking again when it starts transmitting speech.

The point when you start talking is a difficult one for the network. It has to quickly recognize that you have started talking and send the information. Typically, the activity detector takes a few milliseconds to be sure that you really have started speaking. The result of this is that the first syllable that you say is often lost. This is known as *clipping* and is an undesirable side effect of voice activity detection.

Once voice activity detection is working, GSM cannot use the frequencies for anything else. Because you might start talking at any time, and because the gaps between speech are short, the frequencies are of no use for transmission. However, the fact that you are not transmitting means that there is less interference to other users in nearby cells using the same frequency. As a result, it is sometimes possible to reduce the cluster size, especially if frequency hopping is being used, which helps to distribute the reduction in interference to all mobiles in other cells. As an additional benefit, voice activity detection also helps reduce the battery drain since there is less signal being transmitted from the mobile.

CDMA systems implement discontinuous transmission in a slightly different manner. First, they use a more intelligent voice coding system that instead of just being on or off can gradually reduce the information transmitted as the speech stops. Second, the gain from discontinuous transmission is much more direct with CDMA. Because all the users in a cell use the same frequency and because the capacity of the cell is limited by the interference between these users, when users stop transmitting during pauses, this directly reduces the interference to all other users in the same cell, providing an immediate and direct increase in capacity, equal, almost exactly to the percentage of time when discontinuous transmission can be used.

▲

10.3.4   No frequency planning

When different frequencies need to be assigned to neighboring cells, a decision has to be made by the network planners which frequencies to use in which cells. In a CDMA system where each frequency is used in each cell, no such decision needs to be made. Hence, it is true that, in general, CDMA does not require frequency planning although it may require PN code assignment planning. However, this is not a major advantage. Frequency planning can be readily accomplished with today's planning tools and easily adjusted if problems occur. DCA systems do not require frequency planning in any case. Finally, some CDMA systems suggest that frequency planning is performed on a cluster size of 2, for various design reasons, so some frequency planning is required. In summary, this is not a key issue in the selection process for CDMA.

10.3.5   Macrocell/microcell

The concept of using small cells in high-density areas was discussed in detail in Section 8.5. Suffice it to say here that there are situations where smaller cells are deployed within the coverage area of larger cells. This is a problem for CDMA systems. Subscriber units configured for the larger cell will be operating with much higher powers than those configured for the smaller cells. If both cells operated on the same frequency, then the capacity of the smaller cell would be near zero. Hence, different frequencies must be used. Because of the wide bandwidth and hence high capacity of a CDMA system, this may be inefficient, potentially, in the worst case reducing the equivalent CDMA capacity by a factor of two. This reduction does not occur in TDMA since the cells would be assigned different frequencies in any case.

The actual effect of microcells will vary from network to network, but with good planning, capacity reductions of far less than two should be realizable. In summary, this is a problem for CDMA but, although reducing its advantage slightly, is unlikely to change a decision from a CDMA to a TDMA system.

▼   Power control

Another factor that is used to improve the capacity of both TDMA and CDMA systems is power control. This is a simple concept; when the mobile is close to the transmitter and there are no obstructions between them, it can use a lower power because the signal loss is less. This is just the same as needing to shout when you are some way away, but being able to talk more quietly as you come closer. In both systems, the base station measures the power received from the mobile and sends it messages such as "talk louder" or "talk more quietly" depending on what is received.

In both cases, power control reduces interference. In the case of GSM, interference to other cells using the same frequency is reduced, allowing a smaller cluster size in the same way as discontinuous transmission allows a smaller cluster size. In the case of CDMA, the role of power control is much more critical. Because all CDMA signals interfere with each other, if one is received more strongly than the others, then it will interfere more strongly and actually reduce overall capacity. So in GSM, accurate power control can increase capacity, whereas in CDMA, lack of accurate power control can reduce capacity.

For both systems, power control has the added advantage of reducing battery drain increasing the possible talk time.

▲

10.3.6 Risk

TDMA systems have been widely deployed around the globe. CDMA systems are only just starting to emerge, so there is a much higher risk with CDMA that equipment will be delayed, will not provide the promised capacity, or will prove difficult to frequency plan, for example. This risk is continually reducing as experience with CDMA systems grows very rapidly.

10.3.7 Cost

Everything eventually comes down to cost. At the moment, CDMA system components cost more than TDMA system components. However, because of the higher capacity of CDMA systems, fewer base stations are required, resulting in lower equipment bills and lower site and line rental costs. How the two facts balance depends on the actual difference in equipment costs and the extent to which the network is capacity limited. Certainly, in a highly capacity-limited situation, CDMA systems should prove less expensive. Other situations are less clear.

10.3.8 Bandwidth flexibility

CDMA systems can increase the user bandwidth simply by reducing the number of bits in the spreading sequence. TDMA systems can also be bandwidth flexible by assigning more than one TDMA slot per frame to a user. For example, one particular system can assign between 32 Kbits/s and 552 Kbits/s dynamically to one user depending on the load. Hence, both access methods can be made to be approximately equally flexible, although manufacturers may not have designed the capability into individual systems.

10.3.9 Frequency hopping versus direct sequence

CDMA systems come in two variants. The type of CDMA that has been discussed so far is known as *direct sequence CDMA* (DS-CDMA) because the input data is multiplied or *spread* by a sequence or codeword. There is an alternative known as *frequency-hopped CDMA* (FH-CDMA) where the bandwidth of the signal is not directly increased but the transmitter jumps from frequency to frequency. Because more than one frequency is used, the effect appears to be to spread the bandwidth. In practice, only one frequency is being used at one time, so the transmitted bandwidth is not increased, just the spectrum required. The FH is defined as "fast" when the jumps occur more than once in a bit period and "slow" otherwise. At the moment, fast FH is restricted to military applications and is not considered further here.

Readers will remember that FH was already introduced in the context of TDMA systems that move from channel to channel to avoid interference. By some quirk of history or definition, these are exactly the same thing, that is, FH-TDMA and FH-CDMA are identical. It seems far less confusing to use CDMA to mean only DS-CDMA and TDMA to mean TDMA, FH-TDMA, and FH-CDMA and that is the terminology used in this book. Unfortunately, some manufacturers are so keen to be able to place the "CDMA seal of approval" on their equipment that FH-TDMA equipment is badged as CDMA, or even worse as FH-CDMA/TDMA. Readers should assume anything with "FH" in its title is more akin to TDMA than CDMA, although the manufacturer is strictly correct.

10.4 Summary

If the CDMA and TDMA debate was simple it would have been resolved long ago. When trying to understand the issues, the key point to remember is that both are simply ways of dividing up the spectrum. The only real issue is which one can divide up the spectrum with the lowest inefficiency and hence realize the greatest capacity. Initial results seem to show that CDMA is better at this than TDMA and so can provide a greater capacity. Some of the other important points to bear in mind are:

- It is unlikely that there will be any significant range difference between the two systems.

- The advantage that CDMA has associated with not having to frequency plan the system is insignificant.

- CDMA systems perform less well when microcells are deployed.

- CDMA may be less costly but is more risky.

■ CDMA does not have a significant advantage where bandwidth flexibility is concerned.

Making a decision between the two is a very complex issue, typically undertaken by a team of experienced engineers, business planners, and negotiators. To date, where there has been a fair choice between TDMA and CDMA, there has been a slight majority of decisions in favor of CDMA[2]. Perhaps more telling is that it now appears certain that CDMA will be selected as the multiple access method for the next generation of mobile radio systems.

Problem 10.1

What is the key factor around which the TDMA/CDMA debate revolves?

Problem 10.2

What capacity gain do real deployments appear to show that CDMA has over TDMA?

Problem 10.3

Which are the other factors that appear to be of relevance when comparing TDMA and CDMA?

---

2. In many cases the choice is not always "fair." European operators have been mandated to use TDMA (GSM) by their Governments, whereas South Korean operators have been told to use CDMA. In the United States, significant political issues associated with the view that CDMA was "home-grown" while TDMA was imported may have influenced selection decisions.

Part IV

Specific Radio Systems

PREVIOUS PARTS of this book have talked about general principles of radio design and the underlying physics and worked through block diagrams of general systems. This part turns from general radio systems to look at a number of specific examples. There has been no attempt to be comprehensive; this would result in a massive book that would soon be outdated and of little use as an introductory guide. Instead, the GSM system is described in detail and then the most successful cordless, WLL, and satellites systems are briefly examined. By far the most time will be spent on GSM because it is one of the most complicated and most successful systems, and because, since it comes first in this part of the book, once it has been explained, many principles that are the same for other systems also become clear.

11

Cellular Systems

11.1 The difference between analog and digital cellular

Until 1992 all cellular systems were said to be *analog*. This means that the speech signal, which is a continuous smooth analog waveform, as shown in Section 9.2, is encoded directly onto the carrier causing the carrier to vary continually in direct accordance with the speech signal. Since 1992, most newly deployed cellular systems have been *digital*. This means that the speech signal is passed through a speech coder as described in Section 9.2, resulting in a stream of binary digits, or 1s and 0s. This sequence of 1s and 0s is transmitted rather than the original analog speech signal.

Digital systems are generally considered to be better than analog systems. The problem with analog systems is that even a small amount of interference on the radio signal results in a change to the received waveform. The ear is very sensitive to changes, and hence the interference is clearly heard. Another problem is that of *noise*. During propagation, the signal picks up random energy, resulting in a hissing sound in the background. These problems will be familiar to all who listen to AM and FM radio, where the received signal quality is far from perfect, especially when listening on a car radio or portable radio while on the move.

Digital systems can overcome these problems. A digital radio transmits a 1 or a 0. Say it transmitted a 1 but there was some interference during propagation. The receiver might actually receive something between 0 and 1, say 0.8. However, the receiver knows that only 0 or 1 could have been transmitted, so it surmises that a 1 must have been transmitted. It therefore passes a 1 to the speech decoder. By this means, the digital signal can be reconstructed with the effect of noise removed. If the noise was very severe, a 1 might be received as a 0.4. In this case, the receiver would incorrectly guess that a 0 had been transmitted and would generate an error. But the error correction system could detect this and turn it back to a 1. So with a digital signal it is possible to remove noise and interference from the received signal in a manner that is not possible with analog signals.

This difference is entirely akin to the difference between compact disc (which is digital) and vinyl records (which are analog). Most would agree that compact disc is better because there are no scratches, hiss, or other "noise" problems—these are all removed using the mechanisms described previously.

Digital systems also promise higher capacity for the same amount of radio spectrum than analog. There are two reasons for this. The first is that speech coders can "squash" the spectrum required for speech by removing the redundancy as described in Section 9.2. With analog there is no speech coder and so no way of doing this. The second is that with digital, radio frequencies can be reused much closer to the transmitter because the effects of interference can be more easily removed when using digital, by the mechanisms described previously.

These advantages explain why all new mobile radio systems are digital. The reason why earlier radio systems were analog was simply one of technology. Until around 1992, chips capable of performing speech coding and other digital functions fast enough and for a low enough price were not available. Only by the 1990s had technology advanced sufficiently that digital cellular could become a mass market concept. Few would doubt that all future cellular systems will be digital.

Perhaps predictably, not everyone is impressed. Some operators who have switched from analog to digital claim that the analog system was actually preferred by their customers. To some extent this is a rerun of a long debate over whether compact disc or vinyl records were best. It all depends on whether you prefer authentic voice with some noise and interference or voice that has been through a speech coder and subtly altered in the process but has no noise. As with compact disc, as voice coders get better, this argument will be increasingly little heard and the acceptance of digital as the best format for mobile radio systems will become universal. It is also likely that these operators were overlaying digital networks onto a cell structure designed for analog networks and hence not getting the best out of their system.

11.2 Cellular systems in overview

There are a number of different cellular systems in use around the world today. Most cellular systems have a lifetime of around 15 years; hence, at the moment there are analog and digital systems coexisting in many countries. Further, different countries have tended to adopt different systems, with, in particular, Europe, the United States, and Japan all going different ways. Table 11.1 shows the systems currently in existence.

First consider the following list of acronyms, although actually the acronyms mean little and really just act as names for different systems:

- TACS, *Total Access Communication System*;

- NMT, *Nordic Mobile Telephone*;

- AMPS, *Advanced Mobile Phone System*;

- GSM900, *Global System for Mobile Communications*;

- GSM1800, *GSM at 1800 MHz*;

- GSM1900, *GSM at 1900 MHz*;

- DAMPS, *Digital AMPS*;

- PDC, *Personal Digital Cellular*.

Table 11.1
Cellular Systems Worldwide

Name	Analog or digital	Countries of use	Date of introduction	Frequency band	Key features	Comments
NMT	Analog	Scandinavia	Early 1980s	450 MHz and 900 MHz	Allowed roaming between countries	
AMPS	Analog	United States and South America	Mid 1980s	800 MHz		
TACS	Analog	West and South Europe	Mid 1980s	900 MHz	Derived from the AMPS system	Highly successful, will be withdrawn around 2005
GSM900	Digital	Worldwide	1992	900 MHz	See Section 11.3	Developed as a European standard, has become highly successful
GSM1800	Digital	Worldwide	1994	1800 MHz	Same as GSM but at a higher frequency	Originated in the United Kingdom as a means of breaking the cellular duopoly
GSM1900	Digital	United States	1996	1900 MHz	Same as GSM but at a higher frequency	

Table 11.1 (continued)

Name	Analog or digital	Countries of use	Date of introduction	Frequency band	Key features	Comments
DAMPS	Digital	United States	Mid 1990s	800 MHz		Direct replacement for AMPS
CDMAone (previously known as IS-95)	Digital	United States and Asia-Pacific	Late 1990s	800 MHz and 1900 MHz	Only system to use CDMA	See Section 11.4
PDC	Digital	Japan	Mid 1990s	800 MHz		Has not been successful outside Japan

One point of confusion is often the difference between GSM900 and GSM1800. GSM1800 is simply a version of GSM working at a different frequency, there are no other differences. GSM1800 systems work in an manner identical to GSM. Normal GSM mobiles cannot work on GSM1800 systems because they are not designed for operation at the GSM1800 frequencies. However, *dual-band* mobiles are now appearing. These are mobiles that can work on two frequency bands, both the 900-MHz band used by GSM and the 1800-MHz band used by GSM1800. Such mobiles are typically more expensive but have greater flexibility when roaming to other countries. There is a further GSM-based variant called GSM1900, short for GSM at 1900 MHz. Just like GSM1800, this is identical to GSM except that it is in a different frequency band, namely 1900 MHz, because this band is available in the United States for cellular radio. It is expected that there will soon be triple-band phones available, working at 900 MHz, 1800 MHz, and 1900 MHz. These phones will be able to work in all the 120 or more countries using GSM and GSM1800 as well as the United States, something not possible at the moment.

You might well ask why there are so many systems and what the differences are between them. There are two key reasons:

- New digital systems have been designed to replace existing analog systems as the demand for capacity has increased.

- The major countries wanted home-grown systems, and some companies thought that there was money to be made by inventing a better system.

In practice, economies of scale and the need for common standards to allow people to roam from one country to another tend to limit the number of different standards. Broadly, Europe, the United States, and Japan have each developed their own analog and their own digital system, with a couple of analog systems in Europe and a couple of digital systems in the United States. Other countries have tended to adopt the system used by their neighbors. Of all these systems, GSM is becoming the global standard. The reasons for this are many, but probably key is that GSM is the only truly open standard[1] allowing many manufacturers to make equipment to the same standard and competing against each other to drive down prices. GSM is also an extremely well-designed and competent system.

At the very highest level there are few differences between the systems, although at a detailed level the systems can differ dramatically. Broadly all the analog systems are the same as each other and all the digital systems are the same as each other. Some operate on slightly different frequencies than others, but this is only due to the different frequencies that are available in the different countries. Otherwise there is little to choose between them.

With digital systems, manufacturers would like to claim that there are real differences. In practice, all the systems provide nearly identical capabilities. Perhaps the only real difference of note is the capacity, with CDMAone claiming to have higher capacity than the other systems due to its unique use of CDMA. Whether this is true was discussed in Chapter 10.

If this was a more technical text, then the table would include parameters such as the modulation scheme, the multiple access scheme, and the width of the channel. However, knowing all these facts is not

---

1. The United States made the "mistake" of not having any coordinated standards activity during the time that GSM was designed.

really necessary, in the same way that most people who buy cars might be interested in performance and fuel economy but few will care whether they are achieved with 8 or 16 valves, one or two overhead cams, and fuel injection or carburetors. For this reason, these facts have not been detailed here.

Cellular operators need to decide which of these technologies they want to use. They will normally only select one because otherwise their users would not be able to move throughout the country without having two or more mobiles. Actually, most operators do not have a choice. In Europe, the regulators have told the operators that they must use GSM because it is the European standard. In Japan, until recently, only Japanese standards were allowed. It is only in countries like the United States that operators have a free choice. These operators will be balancing capacity against cost and risk in a complex decision. Since capacity, as has been seen, is difficult to compare, risk is subjective, and cost depends more on negotiating skill than equipment choice; then it should not be surprising that there is little consensus of opinion amongst U.S. operators as to what is the best cellular system.

Another point to remember is that it is not always the technically best systems that become the most popular. The classic example of this is the battle of video recorder formats between VHS and Betamax. Although Betamax was widely considered to be technically superior, VHS is now the worldwide standard because it was first into the market and particularly because it became the format for which the most software was available. There is a similar argument for mobile radio. Once one system has become accepted in numerous countries it makes sense for other countries to adopt the same standard because this allows their subscribers to roam, making the service more attractive. Also, the increased sales for this standard provide economies of scale that make the equipment less expensive than other systems. GSM has started down this virtuous route, which often means that even if it has less capacity than CDMAone, it is still preferred for all these other reasons.

In the remainder of this chapter, two of these cellular systems are considered in more detail. These are the digital systems GSM and CDMAone. None of the analog systems have been detailed because these are now slowly being removed. Instead, GSM, the most successful standard, and CDMAone, the most unusual standard, have been explained.

11.3   GSM

By this stage you already know a great deal about GSM because the discussion in the earlier chapters was based around the GSM system. In this section, some of the additional aspects of GSM that have not been covered in earlier chapters are presented. These include the TDMA framing format, the numbering scheme, and the spectrum efficiency of the GSM system.

11.3.1   Framing format

11.3.1.1   Framing format for speech

The GSM framing format is somewhat complex and not essential for a full understanding of the system, so those who are already feeling a little confused might want to skip this part. For those who feel like a challenge, keep reading.

You will remember from Chapter 9 that GSM uses a TDMA framing format, which means that each radio channel is split into slots and each user is given one of eight slots in which to transmit. The slots are numbered 1, 2, 3, 4, 5, 6, 7, 8, 1, 2, 3, 4, …So, a particular user, Peter, might be given slot 1 in which to transmit, which means that he would transmit in slot 1 and then eight frames later would transmit again as the slot structure repeated. Before going too much further, some terminology is required. The frequency channel is normally referred to as a *carrier*. A particular allocation of one in every eight slots is referred to as a *traffic channel*. So each carrier contains eight traffic channels. Each user who is making a call needs their own traffic channel. Each transmission by a user is known as a *burst*. So for the user given all the slots numbered 1, their first burst would be in the 1st slot, their second burst in the 9th slot, and so on. Understanding this terminology is important to avoid confusion. Come back to this paragraph again if things become unclear as the explanation progresses.

This much is no more than would be expected from an understanding of TDMA as given in Chapter 9. The next layer of complexity comes from the need for the mobile to send information back to the network on the signal strength from surrounding cells. Clearly the mobile needs some radio spectrum to send this information. It turns out that the capacity required to send the measurement information is 1/24th of the

quired to send the voice information. The simplest approach
his information is to send 24 bursts containing voice informa-
one burst containing information on the signal strength in
ells. The information in each of the 24 bursts needs to be
tly to cope with the loss of the information that would have
frame. In fact, in GSM the 26th frame is left empty, for
ed with the half-rate speech coding system that will be
This is shown in Figure 11.1, where the abbreviation T
speech (T actually stands for traffic because this struc-
d to carry data as well as speech), S for signaling, and
is kept free.

so when looking at a particular traffic channel (e.g., all the slots 1) the
traffic channel has a structure that repeats every 26th burst. The exact
structure is 12 bursts filled with voice information, one filled with information on the signal strength in the neighboring cells, another 12 filled
with voice information, and then one left spare. Then it starts over
again with 12 slots filled with voice information. Traffic channels are said
to repeat on a 26-burst basis, although it is only the structure that repeats,
not the contents.

When the half-rate speech coder (the advanced speech coder introduced in Section 9.2, which can send speech only using half as much information) is used, only half as many bursts are required for speech.
However, the same capacity is required to send measurement information to the network as with the full-rate coder. Slot 1 is now shared
between two users, say Peter and John, who use the bursts alternately,
as shown in Figure 11.2. The burst that was spare in the full-rate
case (burst 26) is now used as an additional channel to send signaling
information.

Now consider a bit more terminology. Returning to the full-rate
speech coding case, which will be the one used for the remainder of this
section, remember that every 13th burst out of 26 is used to send information about surrounding cells to the network. This information is called
signaling information. In some ways, this 13th burst forms another
smaller channel. A traffic channel was said to comprise one in every eight
slots on the carrier, but if this definition is changed slightly it could be said
that a traffic channel comprises one in every eight slots on the carrier with
the exception of the 13th and 26th burst in every 26 bursts. Another

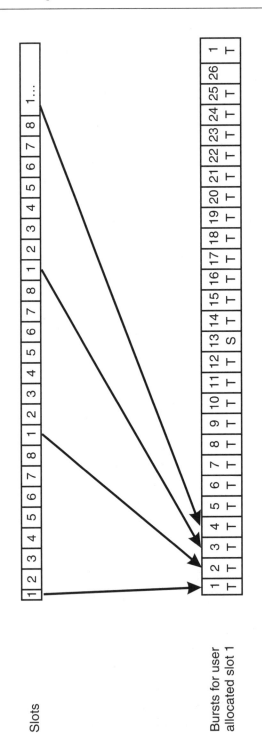

Slots

Bursts for user
allocated slot 1

Figure 11.1   The traffic channel structure for full-rate coding.

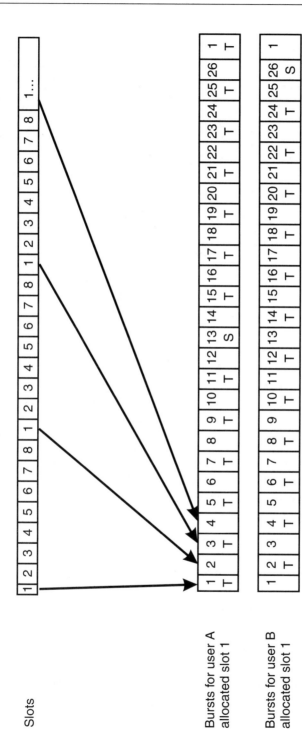

Figure 11.2 The traffic channel structure for half-rate coding.

channel could be defined, called the *control channel*, used to send signaling or control information, which comprises only the 13th burst of every 26 bursts on the traffic channel. This control channel is always used by the mobile allocated the traffic channel using the same slot on the carrier, and so it is called an *associated control channel*. As will be seen later, there are other control channels providing a higher data rate, so this control channel's full name is the *slow associated control channel*. Its purpose is to allow the mobile sending speech information on the traffic channel to send information on the signal strength in surrounding cells.

In summary, on each carrier, or radio frequency, eight mobiles must send their speech and signal strength information. To do this the carrier needs to be divided into eight parts, each of which is then split into two unequal parts. The parts for the speech need to be much larger than the parts for the signaling. Each of these parts might as well be called a channel. This is where many of those trying to understand GSM get confused—when they look into the diagrams trying to find these channels and are unable to do so. The channels are more of a concept than any clear division of the radio frequencies. So the carrier is divided into slots, and each mobile gets one in every eight slots. Each mobile then sets aside one in every 26 of their own bursts for signaling information. That seems straightforward. It is when you are told that this one in 26 bursts is the slow associated control channel that confusion sets in. It is some radio resource reserved for a particular purpose; why not call it a channel?

That is, more or less, all that you need to know about speech channels. Every 26 bursts they repeat their structure, forever, with no deviation from this role.

11.3.1.2 Framing format for control purposes

In earlier chapters, on numerous occasions it was said that the cell site sends information to the mobiles such as the location area identity, the periodicity to use for location updating, and paging messages. Now it is time to look in a little more detail at how it achieves this.

The GSM system has a lot of information to send to the mobiles. However, each piece of information is relatively short. It would be a terrible waste of radio frequencies to use a whole channel to send information on location updating and another channel to send information on paging messages. Instead, they can all be put on the same channel. The only problem with this is that the mobile needs to know where in the channel to find each of the relevant pieces of information. Effectively, the mobile needs a contents list to find its way around the channel.

In fact, the control channel has a number of functions:

- To help the mobile find where the control channels are;

- To provide information as to when the speech and control channel repetition cycle starts;

- To provide information on parameters in the cell;

- To provide information on surrounding cells;

- To provide paging information;

- To allow random access attempts from the mobile.

▼   Getting started in GSM

Just imagine a poor mobile being turned on in a foreign country. It does not know if there are any GSM networks nor which are the control channels, where it should listen for paging messages, where it should send its location update message, and what point the network has reached in its cycle of 26 bursts on the speech channels. There are something like 300 to 400 carriers, each with eight channels on them. How is it going to find the right one? Only with some help from the network, is the simple answer.

The mobile can find out which carriers are being used in the cell just by measuring the signal strength received from each carrier. Those with a suitably high signal strength are being used in the cell. To find out whether a carrier contains a control channel (i.e., one of the eight channels on the carrier, instead of being used for speech has been set aside to provide control information) it listens for a special sequence that the network transmits. This sequence is a pure sine wave (sine waves were introduced in Section 3.2, where it was noted that they were the simplest way to send a changing signal on the radio channel), with no encoded information. Such a sequence would never be sent on a speech channel because there is always information of some sort. The network periodically sends this special sequence on the control channel. The mobile listens to each carrier it finds (at this point the mobile has no idea as to where the slots start and stop, so it cannot pick out individual channels) to see if this sequence occurs. If it does not within the period of time that the network has to repeat this signal, then the mobile moves onto the next carrier.

When it finds this particular sequence, it knows that the network will always send some synchronization information immediately after the sine wave stops. This synchronization information tells the mobile exactly where it is on the control channel cycle and what is happening with the speech cycles. The mobile also knows that the third burst on the control channel will provide it with the contents list for the control channel. From the synchronization information it has found out to what burst it is currently listening, so it waits for the third burst to come round. That tells it what is in all the other slots on the channel, including where the paging messages can be found and where the random access channel is. The mobile can then send a message on the random access channel providing location information. The rest then follows as described in Section 7.2, where roaming was discussed. No wonder that it takes so long for the mobile to find a network when it is turned on in a foreign country.

▲

The control channel has been chosen to repeat on a 51-burst cycle. Almost any number could have been chosen. Interestingly, a bit of mathematics soon shows that $51 = (2 \times 26) - 1$, that is, it repeats with a period exactly one burst less than twice the speech channel. This is deliberate and allows the mobile to measure the signal in surrounding cells more easily, as will be explained later. So, like the speech channel, the structure of the control channel is such that it repeats; however the control channel repeats every 51 bursts while the speech channel structure repeats every 26 bursts.

▼ **How do different length cycles exist on the same carrier**

Now on one carrier, there could be one control channel and seven speech channels (indeed, this is quite a normal configuration). So how can the speech channels repeat every 26 bursts and the control channel every 51 bursts when they are all on the same carrier? Without any problem whatsoever, is the short answer, but since this often confuses, it is worth explaining in a little more detail.

Going back to basics, a carrier is divided into slots. The carrier does not care that one in every eight is given to the same mobile, all it cares is that each slot is only used by a single mobile at any one time. As far as the carrier is concerned each slot is filled with information; it does not care either that every 26th burst for a particular mobile happens to contain signaling information. It certainly would not care if one mobile happened to send signaling information every 26th burst and another mobile chose to send signaling information every 20th burst; as far as it is concerned, all that is happening is that a single mobile is using the channel. Of course, the network needs to know exactly what is happening otherwise it will get confused over what is speech and what is signaling. But that is fine; the network makes up the rules and the mobiles simply do as they are told.

To make it clearer, go back to the basic structure of a carrier. It is just divided into slots. Now, say, the control channel uses the first slot and then there are seven mobiles. The overall structure is shown in Table 11.2.

In this table, round 1 just means slots 1 to 8 on the carrier, round 2 corresponds to slots 9 to 16 and so on. It is clear that slot 1 is following quite a different cycle than the other slots, but so what?

▲

So now it is starting to become clear what a control channel looks like. It actually sends the sine wave once every 10th burst (as shown in Table 11.2) and the sine wave is always followed by synchronization information. So the basic structure of the 51-burst channel becomes "FSxxxxxxxx FSxxxxxxxx FSxxxxxxxx FSxxxxxxxx FSxxxxxxxx F," where F means a sine wave (it is called a frequency correction channel in GSM), S means a synchronization channel, and x just indicates that you have not yet been told what goes in there. It was also mentioned that the contents list always goes in burst 3. There are then only two sorts of bursts left: broadcast information about the cell and paging information. As many bursts as required to provide all the broadcast information are taken from bursts three onward and then the remainder are given over to paging. The contents list (which is also classed as broadcast information)

Table 11.2
How a Carrier Might Look

	Slot 1 (Control channel)	Slot 2 (Speech)	Slot 3 (Speech)	Slot 4 (Speech)	Slot 5 (Speech)	Slot 6 (Speech)	Slot 7 (Speech)	Slot 8 (Speech)
Round 1	Sine wave	Mobile 1	Mobile 2	Mobile 3	Mobile 4	Mobile 5	Mobile 6	Mobile 7
2	Synch	speech	speech	speech	speech	speech	speech	speech
3	Contents	speech	speech	speech	speech	speech	speech	speech
4	Contents	speech	speech	speech	speech	speech	speech	speech
5	Paging	speech	speech	speech	speech	speech	speech	speech
6	Paging	speech	speech	speech	speech	speech	speech	speech
7	Paging	speech	speech	speech	speech	speech	speech	speech
8	Paging	speech	speech	speech	speech	speech	speech	speech
9	Paging	speech	speech	speech	speech	speech	speech	speech
10	Sine wave	speech	speech	speech	speech	speech	speech	speech
11	Synch	speech	speech	speech	speech	speech	speech	speech
12	Paging	signal	signal	signal	signal	signal	signal	signal
13	Paging	speech	speech	speech	speech	speech	speech	signal
14	Paging	speech	speech	speech	speech	speech	speech	speech
15	Paging	speech	speech	speech	speech	speech	speech	speech
16	Paging	speech	speech	speech	speech	speech	speech	speech
17	Paging	speech	speech	speech	speech	speech	speech	speech

Table 11.2 (continued)

	Slot 1 (Control channel)	Slot 2 (Speech)	Slot 3 (Speech)	Slot 4 (Speech)	Slot 5 (Speech)	Slot 6 (Speech)	Slot 7 (Speech)	Slot 8 (Speech)
18	Paging	speech	speech	speech	speech	speech	speech	speech
19	Paging	speech	speech	speech	speech	speech	speech	speech
etc.…	…	…	…	…	…	…	…	…

tells the mobiles where the broadcast information ends and the paging information starts.

So the final structure of a control channel where B means broadcast information channel and P means paging channel might be "FSBBBPPPPP FSPPPPPPPP FSPPPPPPPP FSPPPPPPPP FSPPPPPPPP F."

But where is the random access channel in all of this? To solve this, it is first important to remember that each channel is, in fact, two channels. One is sent from the cell to the mobile, called the *downlink* (because the transmitter is normally higher up on a hill than the mobile and so the signal goes down from the transmitter to the mobile) and the other is sent from the mobile back to the transmitter and is called the *uplink*. The downlink carries the speech from the incoming call to the mobile and the uplink carries the mobile's speech. Without two channels, only one-way conversation would be possible.

In describing the speech channel structure, with its repetition of 26 bursts, it should be remembered that simultaneously a signal is being transmitted from the cell and from the mobile, each on exactly the same 26-burst cycle. Now the control channel is just like a speech channel in that it takes up one channel on the carrier. Just like speech channels, this channel has an uplink and a downlink. However, all the control information described so far is only transmitted on the downlink—it does not make sense for the mobile to send paging information back to the base station. Hence, there is an empty uplink channel corresponding to the control channel downlink. This channel is set aside entirely for random access.

11.3.1.3   Superframes and more

A textbook on GSM would show a complete hierarchy of frames; indeed, if you want to be confused, just take a look at Figure 11.3.

Actually, this is not quite so bad as it looks. You have already come across the TDMA frame (the eight slots, one given to each user) at the bottom of the figure and are familiar with the fact that these slots could be used for a 26 multiframe (for speech) or a 51 multiframe (for control information). Superframes have little meaning. If there is a control channel and some speech channels on the same carrier and they both start at

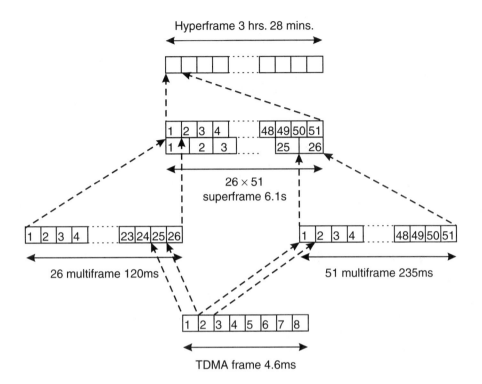

Figure 11.3 Frame hierarchy.

some particular point in time both at burst 1 in their sequence (i.e., the first speech burst for the speech channels and the first frequency correction burst for the control channels), then when do they return to this state? After 26 bursts, the speech channels restart. After 51 bursts the control channels start, but the speech channels are now on burst 25 out of their second cycle. The next time the control channels start again, after 102 frames, the speech channels are on burst 24 of their fourth cycle. They only both start together again after 26 × 51 = 1326 bursts. This is a moderately useful point for synchronization purposes because it helps a mobile switching from a control channel to a speech channel to understand better when to transmit. Each 1326 slots is referred to as a superframe. And that is all there is to it. Hyperframes are even less interesting and not worth further explanation.

11.3.1.4 What goes in a burst

When a mobile gets its chance to transmit, it cannot just send its data. First, as explained in Section 9.8, it needs to ramp up in power at the start of the burst and ramp down in power at the end of the burst if it is to avoid interfering with other mobiles. At some point it also needs to send the sounding sequence, as explained in Section 9.7, so that the cell can understand what ISI is present and set the parameters correctly in its equalizers.

It would seem to make sense to send the sounding sequence first, then the equalizer would understand the reflections in the channel before the speech was transmitted and hence be able to decode the data as it is received. However, as the mobile moves the channel keeps changing and the longer the time period after the sounding sequence has been sent, the more the channel might have changed and the less accurate the parameters used by the equalizer will be, increasing the chance that data is received in error. The way to minimize the time between the data being transmitted and the sounding sequence being transmitted is to put the sounding sequence in the middle of the burst. Then there is a maximum time difference between the data and the sounding sequence of half the burst; whereas if the sounding sequence is at the start, the maximum time difference is a whole burst.

The problem with putting the sounding sequence in the middle is that the first half of the burst cannot be decoded until the sounding sequence is received. This means that the mobile needs to remember the first half of the burst; get the sounding sequence; decode the second half of the burst, which follows immediately after the sounding sequence; then once the transmission has finished, decode the first half of the burst; and reassemble the data.

A typical burst structure is shown in Figure 11.4.

▼ Getting the bursts in the right place—Timing advance

Even though a mobile knows when to transmit from the information sent on the control channel, its signal might not arrive in exactly the right place. This is due to the delay in propagation. By the time the mobile has received the signal that the frame has ended, some time has passed

Content	Tail	Information	Sounding	Information	Tail
N° of bits	3	58	26	58	3

0.577ms

Figure 11.4 Burst structure.

relating to the time taken to get the signal to the mobile. The mobile will then send a signal back and more time will pass as it gets to the base station. The delay will be twice the time taken for the signal to travel to the mobile. Now radio signals travel at 300,000 km/s (the speed of light). If a mobile is say 30 km from the base station then it will take radio signals 0.2 ms to travel to the mobile and back. The burst is only 0.6-ms long, so there is potential to miss the first one-third of the burst and then interfere with the next one-third of the following burst.

The solution to this problem is to advance the timing in the mobile so that before the gap relating to its burst appears at the base station it has already transmitted its information. To do this it needs to know when its burst will appear, which is simple because it understands the slot and frame structure of the system and can predict when its burst will come round again. It also needs to know by how much to advance its timing. It is told this by the base station. The base station listens to when the burst from the mobile is received. If it is received too early or too late, the base station sends a message to the mobile asking it to advance or delay its timing accordingly.

This is fine when the mobile is in a call; however, for the first random access, the mobile will send information with no idea of the timing advance and hence may produce interference of the form discussed previously before the base station can correct its timing advance. There is no real solution to this except to make the random access message shorter than the burst into which it needs to go and for the mobile to aim it somewhere in the middle of the burst. That way, if it drifts somewhat it will still not fall outside its allotted burst and cause interference.

▲

11.3.2   How mobiles measure adjacent cells

One of the things that a mobile has to do while it is in a call is to measure the signal strength in surrounding cells. The use of TDMA makes this relatively straightforward. Say a mobile receives signals from the cell in slot 1. As it happens, it will not be asked to transmit back to the cell at the same time. Transmitting and receiving at the same time requires special components in the mobile that would add cost and complexity to its design. Instead the downlink and the uplink are offset by three slots, so a mobile receives on slot 1 on the downlink and transmits back on slot 1 on the uplink but slot 1 on the uplink occurs at the same time as slot 4 on the downlink.

That still leaves (downlink) slots 2, 3, 5, 6, 7, and 8 for the mobile to make measurements. Actually, this is not as much time as you might think. A mobile needs some time, about one-slot wide, to tune to a different frequency and about one-slot wide to measure a channel. It cannot use slots 2 and 3 to make measurements because it would need slot 2 to tune to the new frequency, slot 3 to make measurements, and slot 4 to retune back to the old frequency, but it should be transmitting to the cell on slot 4. It does have enough time with slots 5, 6, 7, and 8 to make a single measurement.

Mobiles need to do a little more. (Before reading on, make sure you are clear on the difference between slots and bursts—the TDMA channel is divided into 8 slots, a user might be given all the slots 1, each slot 1 is then known as a burst so that burst 1 is slot 1, burst 2 is the second repetition of slot 1, and so on.) Mobiles should try to listen to the control channels in surrounding cells to check that the cell is the one that the mobile thinks it is and to find out the timing in the neighboring cell so that it can rapidly synchronize with it should it need to hand-off. To listen to the control channel it needs to find a frequency correction burst (the pure sine wave used for initial acquisition) and go through the process of acquisition. The mobile is not normally able to do this during slot 6 or 7, which are the only slots on which it can listen to adjacent cells, since there is only a two-in-eight chance that the control channel is on slot 6 or 7 (assuming that the carrier in the adjacent cell is composed of one control channel and seven speech channels) and even less chance that the slots are aligned in both cells, so the mobile can listen to a whole slot rather than half of one and half of the next. However, remember that the mobile also has burst 26, where it does not need to send speech or signaling information. During this time it can listen to eight continuous slots on the neighboring cell, allowing it to listen to one burst of the control channel in the neighboring cells, regardless of which slot the control channel has been assigned.

There is no guarantee that the 26th burst in one cell will align with the time that the frequency correction burst is transmitted. Now the reason why the control channel and the speech channel have different repetition cycles starts to become clear. If they had the same period, it might be that whenever the 26th frame came up in the speech channel in one cell, the say, 8th frame in the control channel in another cell might be present. But because the control channel has a periodicity of twice the speech channel minus one, this means that each time the 26th frame comes up, a different part of the control channel can be read. Because of this difference in periodicity, the mobile is guaranteed to eventually find a frequency correction burst followed by synchronization information and then be able to decode the broadcast information (this process can take up to 30 sec).

11.3.3 Numbering scheme

Up until now, all that has been said about numbering is that each mobile has a phone number. Actually, each mobile has at least three, sometimes more, and in this section the reason for this is discussed. As a minimum a mobile has the phone number you know, the one you tell your friends is your mobile number; this is known as the *mobile station integrated services digital network number*, a nondescriptive name. Most in the GSM community just know it as the MSISDN. However, say this number needs to change. For example, in the United Kingdom, all the mobile numbers are being changed to start with 07 so that people are aware of when they are calling a mobile, which typically results in a high-value call. If this was the number your mobile listened out for and it was changed, then you would need to bring your mobile back to have it reprogrammed. (Well more exactly, you would need to return your SIM card, which is where all the personal details are stored.)

Instead, the mobile is given another number that is only known to the mobile and the network. When someone dials your number, the network looks up what this second number is and sends a message to the mobile using this second number. This is known as the *international mobile subscriber identity* (IMSI), again a name that is not very descriptive. Now if your phone number is changed, someone in the network can just change the database where the dialed numbers (MSISDNs) are converted to internal numbers (IMSI) and your phone does not need to be recalled. Since the internal number is always used to contact your mobile, the phone does not need to know its dialed number. This is why a GSM phone sometimes cannot tell you its dialed phone number!

The internal number is stored on the SIM card so that if you take your SIM card and put it in a different mobile you can still be called by the same number. Because the internal number can be taken out of a phone so easily, it is no good for tracing stolen mobiles. The phone itself also needs what is effectively a serial number, a number that cannot be removed from the mobile—an *international mobile equipment identity* (IMEI).

So, in summary, you have a dialed number that is only used to reach a database within the network where it is converted to an internal number. This is then used to page your mobile. In response, the mobile will let the network know its equipment identity so that the network can check that a stolen mobile is not being used.

You might have more than one dialed number, for example, different numbers for voice, fax, and data calls. Each of these is an MSISDN. In the network these are all converted to the same IMSI, but the network tells the phone whether the call is voice, fax, or data.

11.3.4 Spectrum efficiency

Continually in this book, the lack of radio spectrum has been emphasized and the lengths designers have gone to in order to save it have been explained. Now that the GSM system has been fully explained, it is time to look at just how efficient a use it makes of radio spectrum.

GSM uses TDMA. Each of these TDMA carriers is 200-kHz wide and accommodates eight on-going calls. So in each megahertz of radio spectrum there can be $1000/200 = 5$ carriers giving $5 \times 8 = 40$ on-going calls. Well not quite; for each downlink there needs to be an uplink, so twice as much spectrum is required for 40 on-going calls. This is often written as 2×1 MHz to show that two chunks of spectrum, each 1-MHz wide, are required, one for the downlink and one for the uplink. Further, at least one channel is needed for control purposes, so only 39 or perhaps even 38 on-going calls can be handled, but the control channel capacity is normally neglected in calculations because it is not greatly significant.

In Chapter 3, it was explained that because the same frequency cannot be used in neighboring cells, only a fraction of the available spectrum can be used in any one cell. In a typical GSM network the cluster size, as defined in Section 3.3, is typically 12. This means that of every megahertz, only 1/12 can be used in any cell. So, actually, in each cell, in each 2×1 MHz, $40/12 = 3.3$ simultaneous calls can be accommodated. This is the spectrum efficiency of GSM, 3.3 calls per 2×1 MHz per cell.

In Section 9.2 it was mentioned that there is a half-rate GSM speech coder. If this is used, each voice call only takes up half as much spectrum, so twice as many voice calls per megahertz can be supported. If the half-rate coder is used, then the spectrum efficiency of GSM rises to 6.6 calls per 2×1 MHz per cell.

If frequency hopping is used, then it is sometimes possible to reduce the cluster size to a lower level. Some manufacturers have even claimed that the cluster size can be reduced to as little as 3. There is little evidence that such small clusters can actually be achieved in real life, but certainly there is some scope for reducing the cluster size, perhaps to 9 or 7, and hence increasing the capacity by around 30%.

11.4   CDMAone

CDMAone is the new name for a radio system that was previously known as IS-95, or Qualcomm CDMA. Much of CDMAone is the same as GSM. For example, the overall architecture, the use of paging, location updating, random access, and security, is all the same as it is for GSM. The only real difference lies in the detailed parameters of the radio signal sent between cell and mobile, the sort of parameters discussed in Part Three. The biggest difference is that CDMAone uses the CDMA multiple access method, whereas GSM uses the TDMA multiple access method, both of which were discussed at length in Chapters 9 and 10. Since this book is not overly concerned with the very detailed modulation and transmission parameters that are only really necessary for mobile radio engineers to understand and the differences between CDMA and TDMA have already been examined in Chapter 9, this section will be very short. As you will have already gathered from Chapter 9, CDMA is very complicated, so even a brief look at the CDMAone system is bound to be complicated.

The intention here is not to describe CDMAone in as much detail as GSM. GSM was described in detail so that you could understand the workings of a representative cellular system. Providing details on the CDMAone framing structure would not add to your understanding of how cellular systems work. Instead, they key differences of CDMAone from GSM are explained here.

Before getting into the complexity, it is worth summarizing the key issues. Nearly all of CDMAone is the same as GSM. What is different is that it achieves a different spectrum efficiency than GSM due to its different radio parameters. The only other differences are a different voice coder, which may be better than GSM's, giving a better speech quality, and a different hand-off method, which may give better quality; and that is it. When deciding whether to use GSM or CDMAone, the key concern of the operators are which will give them the best spectrum efficiency, allowing them to have the highest possible number of subscribers in their limited spectrum assignment (and, of course, which manufacturer will make them the best offer).

A description of a generic CDMA system was provided in Chapter 9. The only additional information needed to understand the CDMAone system is the actual parameters that are used. These are detailed here but do not make for very interesting or enlightening reading. Mostly, it will be sufficient to understand the basic principles of CDMA and that CDMAone is simply a particular implementation of these basic parameters.

The system uses a cluster size of 1, where each cell transmits its CDMA signal on the same carrier. This is possible in CDMA because the system works by extracting signals from interference; hence, if there is interference from neighboring cells as well as within the cells, this does not cause undue problems. In fact, interference from neighboring cells reduces the capacity in the cell under consideration. However, by using a cluster size of 1, more frequencies are available in each cell. The number of additional frequencies available outweighs the loss in capacity on any particular frequency, maximizing the overall system capacity.

The downlink transmission consists of a permanent signal and a number of radio channels. The permanent signal is called a *pilot tone* and corresponds to the control channel in the GSM system. It is used by the mobile to estimate the pathloss, so as to set power control initially, and to acquire synchronization to the network. Other channels are set aside for paging and other downlink information. Speech is encoded using a similar voice coder to that used in GSM, and then error correction coding is applied to generate a data rate of 19.2 Kbits/s. This is then spread or multiplied by the CDMA code. In this system the codes are selected from the family of 64-bit Walsh codes (remember that Walsh codes are the most common type of orthogonal codes used in CDMA systems). The multiplication of 19.2 Kbits/s by 64 results in a 1.228-Mbits/s signal, which is then transmitted.

The uplink transmission is slightly different. Speech is generated in the same manner, but more powerful error correction coding is being used to give a bit stream of 28.8 Kbits/s (as opposed to the 19.2 Kbits/s used on the downlink). This signal is then spread in a more complex manner than the uplink using the Walsh codes in a slightly different manner, resulting in an intermediate data rate of 307 Kbits/s. This does not provide enough spreading to counteract the interference and so each mobile generates a unique code to complete the spreading. This unique code is based on a different code family from Walsh codes, the family called the PN codes. This multiplication further spreads the intermediate data by a factor of 4, resulting in a data rate of 1.228 Mbits/s, the same as the transmitted rate.

The reason why the uplink is treated in a different way from the downlink is because of synchronization. On the downlink all the signals are synchronized because they are all transmitted from one point, the base station. On the uplink, all the signals are transmitted by different mobiles, each at a different distance from the base station, and hence they do not all arrive at exactly the same time. Recall from Section 9.8 that if Walsh codes are not synchronized they lose their property of orthogonality. The use of a combination of Walsh codes and PN codes seems to overcome this loss of synchronization better. It should also be mentioned that the reasons for some of the design decisions taken in the CDMAone system are far from clear!

Calculating the capacity of a CDMA system is extremely difficult. As explained in Chapter 10 some approximations can be made. However, perhaps more useful are the results from real deployments where up to 15 voice channels per carrier were employed. Since one carrier is 1.228-MHz wide, the capacity per cell per megahertz is given by approximately 12 voice calls per cell per megahertz. Comparing this with GSM, recall that GSM has 8 voice channels per carrier and 5 carriers per megahertz. Hence, there are 40 voice channels per megahertz, but with a cluster size of 12, this reduces to $40/12 = 3.3$ voice calls per cell per megahertz. So basic GSM has only around 25% of the capacity of a CDMA system. If, however, the full GSM armory is deployed including frequency hopping and discontinuous transmission, significant reductions in the cluster size are possible, allowing the cluster size to fall to around 7. With a cluster size of 7, GSM achieved $40/7 = 6$ voice calls per cell per megahertz. This is around 50% less than CDMA, but when other factors that reduce CDMA capacity, such as soft handover, are taken into account the capacity gain may be only around 30%.

CDMA has a key advantage over TDMA in that when a new sector is added, the capacity is increased by something approaching an extra 15 channels. The interference is increased, because there are more mobiles using the same frequency in the same area, but the increase is not as much as if sectorization were not used because the directional antennas at the base station ensure that only a limited signal leaks into the other sectors. Hence, although there is a loss of capacity in each sector (perhaps the sector capacity falls to two-thirds), there are now three sectors and hence the overall capacity is increased by a factor of 2. In TDMA, the addition of new sectors only provides limited capacity gains (as discussed in Chapter 8). This makes it difficult to perform a strict comparison between the different technologies. Here, a single sector per cell has been assumed, but it is noted that the CDMA system can be enhanced more readily than the TDMA system. For this reason, the estimate that CDMA is actually 30% better than GSM has been used in this book. This is probably conservative; time will tell whether CDMA can actually outperform GSM by more than this.

11.4.1 Soft hand-off

Because CDMA transmissions in each cell are on the same frequency, mobiles are able to receive transmissions from a number of cells at the same time. Typically, they reject all but the transmissions from the wanted cell in order to decode the signal sent to them. However, the ability to listen to more than one cell at the same time proves useful when the mobile is handing off from one cell to another. During hand-off the network could send the same signal to the mobile from two different cells. The mobile could listen to the signal from both cells while it is passing between them and, when well within the new cell, could move back to listening only to the new cell. Such an approach minimizes the risk that a mobile will be lost during hand-off, prevents any short break in speech while the hand-off is being performed, and maximizes the voice quality during a time when the mobile is on the edge of a cell and might otherwise expect a poor-quality signal.

However, all these gains do not come for free. On the network side, two radio channels are required when a mobile is in the process of hand-off, reducing the overall system capacity, some estimate by as much as 40%, depending on the size of cell, the speed of the mobile, and the overlap between the cells. On the mobile side, although the mobile only needs one decoder to take the signal from the radio frequency to the voice frequency, it needs two subsequent decoders to monitor each of the two signal paths, adding to the complexity of the mobile. Nevertheless, the gains are thought to be worth these problems and soft hand-off is a feature available within CDMAone.

Problem 11.1

Why is the 26th burst on a full-rate speech channel kept free in GSM?

Problem 11.2

What different types of information are sent on a GSM control channel and what is the purpose of each of these bits of information?

Problem 11.3

Why is the sounding sequence put in the center of the burst?

Problem 11.4

What cluster sizes are used by GSM and by CDMAone? Why are they different?

Problem 11.5

What is soft hand-off?

12

Cordless Systems

12.1 Overview of cordless

Cordless systems are those phones you buy in a shop, plug the base unit into the phone socket in your wall, and then walk around your house with the radio handset. What you have actually bought is a combined cordless home base station and handportable allowing you to take calls coming into the fixed line using a cordless phone. This was the original cordless concept, and you can still find cordless phones in most phone shops. The combined home base station and phone normally sell for around $100 for an analog phone and $200 for a digital phone.

A number of important design decisions arose from such origins. First, the range of the phone from the base station only needed to be around 200m and so low-power designs were used. Second, the phone and the base station needed to be inexpensive, so the phone design was simple, avoiding complicated speech coders and channel equalizers, for example. Third, frequency planning was not possible because the users owned the base station as well as the mobile and so the phones needed to seek a low-interference channel whenever they were used. All the technologies detailed in this chapter have these fundamentals as part of their design.

Other applications for cordless phones were soon developed. A key application is the replacement of an office PABX (the switchboard in the office that directs incoming calls and allows internal calls to be made within the office) and wired phones with a central cordless phone system able to provide mobility and appropriate exchange functions. Systems designed for this application typically use TDMA to allow the central PABX to use as few amplifiers as possible. *Digital Enhanced Cordless Telephone* (DECT), which will be explained in the next section, is an example of such a system.

The final application for which cordless has been adopted is known under a wide variety of names, most often *telepoint*. Telepoint is the deployment of cordless base stations in streets and public areas to provide a low-cost service that, at first, appears similar to cellular. The key difference is that the base station range is so low that coverage can only be provided in high-density areas, unlike cellular where more ubiquitous coverage is provided. Telepoint systems have broadly failed. The United Kingdom launched four networks that failed, as did networks in France, Germany, and a variety of other countries. Typically, the lack of coverage compared to cellular makes the service unattractive. Telepoint has been successful in a very limited number of places, namely Singapore, Hong Kong, and more recently Tokyo. These are all very high density cities where the populace rarely leaves the city areas. In such cities, widespread deployment of cordless base stations can be worthwhile because of the high user density and the relatively good coverage that can be provided. The high capacity that results from using very small cells (see Chapter 8) allows many more users onto the network than is possible for a cellular system, allowing call charges to be lower than cellular, making the service an attractive proposition. It is unlikely that there are many other cities in the world where cordless will be successful in a telepoint application and its main sales will remain as home and office phones.

More recently, cordless technologies have been offered for WLL (which will be described in the next chapter). From the preceding description it is apparent that the key characteristics of cordless will be low range and high capacity. In a WLL deployment it is possible to extend the range using directional antennas. The technology is relatively inexpensive and any need to deploy additional base stations need not incur a cost penalty over other radio systems.

The key cordless technologies of DECT and *Personal Handiphone System* (PHS), which is the Japanese cordless standard, are now described in more detail.

12.2 DECT

DECT was standardized within Europe during the early 1990s, initially for use as a wireless office PABX. It has recently also been suggested for deployment in WLL networks as well as cordless applications. Its key strengths are good voice quality; the ability to provide a high bandwidth to the user, allowing them to send large computer files quickly; and the avoidance of frequency planning.

DECT has an advantage over many of the other cordless and cellular technologies, namely, the availability of unlicensed and standardized spectrum. In most countries, especially in Europe, the 1.88- to 1.9-GHz frequency band has been set aside for DECT. Therefore, operators can easily gain access to spectrum and manufacturers are able to achieve good economies of scale with their equipment. Using the *generic access protocol* (GAP), currently being developed, subscriber equipment from one manufacturer should be able to interwork with network equipment from a different manufacturer, easing concerns over multisourcing.

DECT transmits using TDMA in a similar manner to GSM. Radio channels are spaced 2 MHz apart. Each radio channel is divided into 24 timeslots. Nominally, 12 are for base station-to-subscriber transmission (downlink) and 12 for subscriber-to-base station transmission (uplink), but this can be varied dynamically if there is more information to be transmitted in one direction than the other.

Because it transmits in both directions on the same frequency, DECT is said to be *time division duplex* (TDD). Literally, the time is divided between the different duplex directions. TDD allows simpler mobile design. Importantly, the fact that the same radio channel is used in both uplink and downlink directions also allows the use of antenna diversity, a technique discussed subsequently.

▼ Diversity

As mentioned when discussing one of the advantages of frequency hopping in Section 9.6, the fast fading pattern changes as the frequency (and hence the wavelength) changes. It is quite possible for a mobile receiving on one frequency and transmitting on another to have the downlink in a fade but the uplink transmitting a signal that is received strongly at the base station. One way of overcoming fades is antenna diversity. In this arrangement the receiver has two antennas positioned ideally at least a wavelength apart (but not an exact multiple of wavelengths). The hope is that if one antenna is in a fade, then the other one is not. By taking the signal from the antenna with the strongest signal during each burst, significant improvements in tha received error rate can be achieved. This is known as spatial diversity. Unfortunately, the use of diversity antennas on the subscriber unit is often difficult. Because the unit is relatively small, it can be difficult to space the antennas far enough apart and the additional cost of a second antenna can often make the equipment uncompetitive. TDD provides a solution to this. When receiving a transmission, the base station decides which of its diversity antennas is receiving the strongest signal. If it also transmits from this antenna, then the subscriber unit will receive a good quality signal. Hence, a diversity path is achieved to the mobile without the need for a second antenna in the mobile.

▲

Within each of the 24 slots on the TDMA frame a 32-Kbits/s bearer capability is provided. This allows ADPCM speech or a range of data options. Slots can be concatenated to provide up to 552 Kbits/s per user (requiring 18 of the 24 slots) if needed (although this will, of course, prevent many other users from accessing the base station at the same time).

DECT also has DCA. This is where the base station listens to all the frequencies available to the network and finds the one with the lowest interference from surrounding cells. It allows DECT systems to exist without the need for frequency planning and to maximize the system capacity in cases where the interference from surrounding cells is relatively low.

The high data rate of DECT can cause some problems with ISI. DECT does not have an equalizer but transmits data at a relatively high rate (around 1.1 Mbits/s). Any reflected paths with an extra path length of 260m more than the main path can cause problematic ISI. When an omni-directional antenna is used in an urban environment, the ISI problem has been found in practice to limit the range of DECT to under 300m regardless of the power that is transmitted.

One key advantage of DECT is the simplicity of the system, resulting in relatively inexpensive equipment costs. Because highly complex radio transmission methods and protocols have been avoided, DECT base stations are available at only a fraction of the cost of cellular base stations.

12.3 PHS

The PHS is a Japanese standard for use in the 1895- to 1918-MHz frequency band. It has proved highly successful in telepoint applications in Japan, where the subscriber density is very high and is generating much interest in Asia-Pacific countries.

Like DECT it is based on a TDMA-TDD approach. It uses a carrier spacing of 300 kHz providing four channels, although typically one of these will need to be used for control information. It uses 32-Kbits/s ADPCM speech coding. Also like DECT it uses DCA to optimally select the best frequency. As can be seen from this short description, PHS is extremely similar to DECT in most respects. Its modulation is slightly more advanced, enabling it to achieve a slightly better spectrum efficiency, but broadly that is all. Hence the reason why PHS is not described in great detail—an understanding of DECT is sufficient to understand PHS[1].

Recently there has been much publicity about PHS. Sales in Japan have been very successful and the Japanese have been trying hard to sell

---

1. This leads to the question "Why did the Japanese invent PHS?" Broadly, the answer to this is to boost the success of Japanese industry since DECT manufacture was already dominated by European companies.

their technology overseas. The PHS system is a perfectly adequate cordless system but does not offer any great advantages over DECT and is certainly only suitable for cordless applications. It has been doing well in Japan due to both the very high density of Tokyo, making the deployment of cordless an economic venture; and the very aggressive way that it has been sold, with few, if any of the operators or manufacturers actually currently making any money out of PHS. Because of the dramatic increase in sales, many observers outside Japan have started to think that PHS must have some advanced characteristics that render it better than other phone systems. This is not true, as explained in more detail next. PHS is simply another cordless standard, similar to DECT, that happens to have been aggressively marketed in Japan. It is most certainly not, as one PHS supporter recently declared, "a significant step on the way to the next generation cellular phone systems."

There has been much discussion as to whether DECT or PHS is best, often with the key arguments being lost in the battle between two parties, both of whom have significant commercial interest at stake. It is instructive to examine some of the comparisons in order to understand better how to select between different systems.

According to Ericsson, the key differences are:

- DECT has a 30% greater capacity than PHS.

- DECT has a reduced infrastructure cost compared to PHS.

- DECT is easier to plan than PHS.

- DECT can provide 64 Kbits/s and ISDN bearers whereas PHS cannot.

- DECT can provide repeaters whereas PHS cannot.

- DECT can interwork with GSM whereas PHS cannot.

- DECT systems exist better in an uncontrolled environment than PHS.

According to Fujitsu the key differences are:

- PHS has a lower delay than DECT.

- PHS has a reduced infrastructure cost compared to DECT.

- PHS has a greater capacity than DECT.

- PHS has simpler connection to the PSTN than DECT.

- PHS base stations have lower power consumption than DECT base stations.

Claims of this sort are quite normal in the world of cellular and cordless systems. A comprehensive comparison between two systems is complex because of all the variables involved and so manufacturers rely on operators being unable to do this comparison for themselves and hope to influence them with claims. Typically, the claims have some truth but often their system has been presented in the most promising light whereas the other system has been viewed pessimistically. Sorting out all these claims keeps consultants busy for months.

Most of the claims are actually true, although the impact of items such as the simplicity of connection to the PSTN is minimal. Clearly, though, both cannot be right on the questions of capacity and cost since they both claim their system is the best. Examining the capacity issue in more detail and taking all relevant factors into account, calculations show that the PHS capacity is typically somewhat greater than the DECT capacity.

The question of infrastructure cost is more difficult to evaluate until the actual costs of the different base stations are known. DECT proponents contend that fewer DECT transceivers are required because each transceiver provides 12 channels compared to the 4 of PHS. Ericsson claims that a DECT base station will cost 1.1 times a PHS base station but provide 3 times the capacity, hence the overall reduction in infrastructure costs. Most likely, the base station costs will be dominated by the economies of scale achieved and not the manufacturing difficulties and, in this area, it remains to be seen which technology will be the most cost effective.

There is a difference between the services that DECT and PHS can provide. Because more than one slot can be given to the same user, DECT is much better able to provide high bandwidth and multiple line per subscriber services than PHS. Given that the costs are probably not too dissimilar, it is this improved functionality that gives DECT the edge over PHS.

As a final point of note, the PHS spectrum allocation is not assured in European countries. Hence, it may prove more difficult to deploy PHS than DECT from a spectrum allocation viewpoint in all but the Asia-Pacific countries. In the United States, both DECT and PHS spectrum allocations are used for cellular applications, so neither are likely to enter into widespread use.

Problem 12.1
Why have telepoint systems failed in most countries?

Problem 12.2
What is the key advantage of TDD?

Problem 12.3
Why can DECT achieve a much greater range when deployed as a WLL system than when deployed as a cordless system?

Problem 12.4
What are the key differences between DECT and PHS?

Problem 12.5
Why is no sounding sequence sent with cordless systems unlike cellular systems?

13

Overview of Wireless Local Loop Systems

13.1 Introduction

Most people are familiar with cellular radio; most readers will have used a cellular phone at some time in their life. *Wireless local loop* (WLL), which is the use of radio technology to provide a telephone connection to the home, is a less well known concept. WLL systems have a very similar "architecture" and similar problems to cellular systems. Before explaining the similarities between the two systems, this chapter provides some background information on WLL systems and why they are of interest.

Recent predictions from a number of key telecommunications analysts have shown that the WLL market is set for dramatic growth in the next few years. Perhaps most startling are the predictions that by 2002 the number of wireless lines installed worldwide will exceed the number of wired lines. Despite the dramatic predictions of growth, WLL remains a little understood topic, often confused with cellular or cordless telephony.

Historically, the way that homes were connected into the telephone network was using copper cable buried in the ground or carried on overhead pylons. This connection is known as the *local loop*—*loop* because each connection requires two wires that look like a loop when drawn on a diagram and *local* to distinguish the connections to homes from connections made between different switches. WLL replaces the local loop section with a radio path rather than a copper cable. It is concerned only with the connection from the distribution point (typically a junction box alongside the street connecting subscribers within a radius of 1 km) to the house; all other parts of the network are as for a traditional wired arrangement. In a WLL system, the distribution point is connected to a radio transmitter/receiver, a radio transmitter/receiver is mounted on the side of the house, and a cable is run down the side of the house to a socket inside the house. This is a socket identical to the one into which users currently plug their home telephones. Hence, apart from a small transmitter/receiver on the side of their house, the home subscriber does not notice any difference.

Using radio rather than copper cable brings a number of advantages. It is less expensive to install a radio than to dig up the road; it takes less time; and radio units are only installed on a house when the subscribers want the service, unlike copper, which is installed when the house is built. The latter point reduces considerably the "upfront" investment required before revenue is received from subscribers, which is important in reducing risk and overall financing requirements. These advantages will be considered in more detail in the remainder of this chapter.

An overview of a typical WLL network architecture is shown in Figure 13.1, where its similarity with a cellular network is obvious. Broadly, the network consists of three key elements: the base stations, the switch, and the subscriber units. There is also a network management system which monitors all functions of the network.

WLL systems are being deployed in a range of different locations, often characterized as developing countries, Eastern European countries, and developed countries.

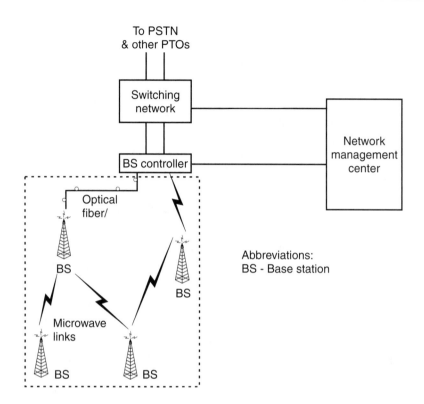

Figure 13.1 General network configuration.

- In developing countries there is little telephone infrastructure and WLL is seen as the most cost-effective way to provide voice telephony to those who currently do not have access to the PSTN. The major difficulty is often whether the local population can afford to pay for telephony.

- In Eastern European countries there is often a long waiting list for telephony and WLL is seen as the most rapid way to provide basic telephony, with perhaps limited data capabilities, to overcome the waiting list.

- In developed countries operators are attempting to provide competition to the existing wired operator and see WLL as the lowest cost means of providing that alternative. Here, low cost coupled with good data capabilities are seen as essential in order to take market share from the incumbent operator.

13.2   Access technologies

An *access technology* is a means of connecting a home into the PSTN, or providing it with access to the PSTN. Until recently, there was only one access technology, namely, the copper wiring that links most homes in the developed world into the telephone network. More recently, other potential access technologies have been devised. If you are involved in WLL services it is important to understand what access technologies are available since these will form the competition to WLL and will have a significant effect on the profitability of the network.

It is important to differentiate between *bearers* and access technologies. The copper wire or radio path is a basic bearer, whereas a particular computer sending a signal over that bearer might form the access technology. Hence, there are typically more access technologies than there are bearers. The key bearers are the copper cables belonging to the telephone network, the cables belonging to the cable TV networks, mobile networks, and WLL networks.

It is with the copper cables that there is the widest range of different technologies able to use the bearer, from basic connections over which modems can provide a moderate data rate, through so-called ISDN lines that provide a medium data rate and will be explained later, to the high data rate technologies proposed such as *Asymmetric Digital Subscriber Lines* (ADSL). ADSL is a complicated means of getting much higher data rates down telephone lines—being a wire-based technology it falls outside the scope of this book. Such a high-speed technology would allow the national telecommunications company to deliver services such as video on demand through existing connections. ADSL is currently under development and it is still far from clear whether it will be successful.

Before going any further, consider a bit more terminology. In the previous paragraph the national telecommunications company was mentioned. This means companies such as BT or the regional Bell operating companies, the companies who currently carry the majority of the telephone traffic in a country and who own the copper wires connecting homes into the PSTN. These companies are often referred to as the *Post and Telecommunication Organizations* (PTOs) because most of them also managed the postal service at some time. They are also referred to as the "incumbents," because as far as new operators are concerned, the PTOs have already gained a large share of the market and have an entrenched position.

Cable TV networks are being upgraded to provide high bit rates per subscriber and form a keen competitor to other access networks. Cellular radio is also a contender for the local loop, although recent initiatives by cellular operators to encourage subscribers to use a mobile phone in place of their fixed phone have generally failed due to poor voice quality and high costs. However, as the cellular market becomes increasingly competitive, more such initiatives might emerge.

Another alternative access technology, closely aligned to WLL, is the video distribution system that is available in a range of variants known as *microwave video distribution system* (MVDS), *microwave multipoint distribution system* (MMDS), and *local multipoint distribution system* (LMDS). These systems are widely deployed in the United States but less so elsewhere and provide 50 or so TV channels to a small aerial mounted on the side of the house—they differ only in the frequency at which they operate and whether they use analogue or digital transmission. It is worth explaining how these differ from WLL. At the overview level there is no difference; both use ground-based radio systems to send signals to receivers mounted on the side of houses. The differences lie in the services each system is tailored to provide. Video distribution systems are designed to provide numerous video signals from the transmitter to the viewer in a broadcast format, with limited capabilities for transmission from the viewer back to the network. WLL systems are designed to provide telephony and data services with an equal capacity both to and from the network. Broadly, video distribution systems do not provide telephony well, or economically, and WLL systems do not provide video broadcast capabilities. Because WLL systems do not require the wide bandwidths needed for video distribution, they can be accommodated at lower radio frequencies with resulting greater range, less-expensive subscriber equipment, and hence lower cost. If video distribution is not required, WLL currently provides a less-expensive and better quality telephone service than video distribution systems.

Table 13.1 provides a summary of all the access technologies currently being deployed.

It is clearly apparent from this table that WLL is not competitive in terms of providing high bit rates to subscribers. The key role fulfilled by WLL is where there is a limited requirement for high data rates—or, more precisely, where most users are not prepared to pay more for higher data rates than they are currently using. It is as a relatively low data rate bearer that WLL systems are being deployed around the world, with

Table 13.1

Comparison of the Different Access Technologies

Access technology	Data rates	Advantages	Disadvantages
Ordinary telephone line	Low	Low-cost, immediate installation	Blocks telephone line, relatively low data rates
ADSL	High	Can provide applications such as video and home shopping on existing lines	Unproven, expensive, and will not work on all lines
Cable line	Very high	Relatively cheap, allows subscribers to receive both TV and telephony	Cable penetration lower than copper, only a few simultaneous users can be supported
Mobile radio	Mostly very low	Can be used immediately where coverage	Limited data capabilities, cost, lack of coverage
Telephony WLL	Medium	Economic to provide, reasonable data rates	High data rates not possible, new infrastructure required
Video WLL	Very high to the user, low to the network	High data rates at low cost	Low data rates to the network, short range

subscribers being offered lower tariffs than those of the existing fixed line operator.

It is unlikely that WLL will develop sufficiently high bit rates to provide video distribution or to compete with the ADSL technology; the high data rate market will either be supplied by video distribution systems or ADSL depending on how the relative costs and reliability emerges. However, it is easy to underrate WLL because of its inability to provide high bit rates—in the same manner that the domestic saloon car could be underrated due to its inability to achieve a 200 mph top speed. In practice, few subscribers will be prepared to pay for high-data-rate access, leaving WLL well placed to address the majority of the local loop marketplace. The requirements for high data rates is a topic to which we return in Chapter 15.

Some advocates of WLL are inclined to promote its apparent capability to provide mobility as well as a connection into the PSTN. This apparent capability arises, as will be seen, from the fact that many WLL technologies are based on cellular radio standards. Such a view is mistaken. To be economic, WLL systems need to offer a service of equivalent voice quality to fixed line at a sufficient cost reduction to persuade subscribers to migrate from their current fixed line to a wireless line. The provision of a low-cost, high-quality service is achieved through roof-mounted directional antennas at the subscriber's premises, pointing toward the base station and hence providing a huge gain in signal strength compared to that which would be received inside the house using a cellular phone. Without such a gain, of the order of 100 times more cells would be required to provide a similar quality of service, rendering the WLL network uneconomic.

13.3 The attraction of WLL systems

The key attraction of WLL is simple—it can provide an equivalent local loop connection to copper cable at a lower cost. It does not necessarily provide a better or different service, simply one that is less expensive, allowing the new operator using WLL to compete on price against the incumbent. The economics of fixed versus wireless are relatively simple to understand. The key points to note are that:

- The cost of installing wired systems is broadly dependent on the cost of labor (to dig up the roads), which tends to rise at a rate greater than inflation. The cost of wireless depends broadly on the cost of the subscriber units, which tends to fall over time with increasing economies of scale.

- The cost of wired systems is critically dependent on the distance between houses and the penetration levels achieved, whereas the cost of wireless is broadly independent of these factors.

- If a subscriber moves to a different operator, in the case of wired access the investment is lost whereas in the case of wireless the subscriber unit is typically removed and installed elsewhere.

- The cost of wired systems is incurred prior to marketing to the users, whereas much of the cost of wireless is not incurred until the users subscribe to the network.

In the past, the cost of the electronics required for WLL has been greater than the cost of laying cables, hence the reason why most telephone lines are wired rather than wireless. Only in the last few years, with the economies of scale generated by cellular, have prices fallen to levels where WLL has become economic, with further falls likely in coming years. This will result in an increasing trend toward WLL as the predominant means of providing subscriber connection into the PSTN.

The relative costs can be illustrated by the following example, which uses approximate 1997 cost levels. First, cabled costs can be estimated as follows.

Suppose that houses in a particular area are 10m apart and have front gardens that are 5m long. Suppose that the cost of cabling in the street is $30/m while in gardens it is $20/m. Then suppose that termination costs are $50 per home.

The cost per home in the case of 100% penetration is then given by 10m × $30/m = $300 plus 5m × $20/m = $100 plus the termination costs of $50. This equals $450/home.

Now imagine that the penetration drops to 20% (a figure typically achieved by cable operators). Now it is necessary to pass five homes before a cable can be installed. The cost to get the cable to the road outside the house is now $5 \times 10m \times \$30/m = \$1,500$. The other costs remain the same, resulting in a total cost of $1,650 per house.

Maintenance cost on buried cables is relatively high. Although they are underground they can suffer from flooding and from disruption by others who dig up the streets, such as gas and water utilities. When problems occur they often require that a stretch of street is redug in order to locate the fault and replace the cable. Maintenance costs on buried cable are typically some 5% of the installation cost per year.

By comparison, the equivalent costs for WLL can be found as follows.

Suppose that there are 5,000 houses in a cell of radius 3 km, that the transmitter costs $150,000 to buy and install, and subscriber units are $800 each including installation. Then the fraction of the transmitter cost applying to each house is $30 and the total cost per house is $830. This is about twice the $450 per house cost calculated in the cable case.

Now suppose that penetration is only 20%. The transmitter cost per subscriber then rises to $150 per house and the total cost to $950. This is significantly less expensive than the $1,650 calculated in the cable case.

A key difficulty here is comparing the number of houses in a certain radius with typical distances between houses. Clearly, significant variability can occur in this area.

Figure 13.2 shows the relative costs of cable access and wireless access for a medium density environment. The advantages of WLL where penetration is low are clearly obvious. Simply, this shows the key roles of WLL.

The figure shows WLL providing a less-expensive access technology than cable for low-penetration levels. As wireless system costs fall in coming years, any difference is likely to increase. Once the relative maintenance costs are included, these differences become even greater. In most cases operators will be working at low penetration levels as they compete with the existing telephone providers.

There are further issues in some countries. For example, in South Africa, buried copper cable tends to get dug up and turned into copper bracelets for sale in the local market. In tropical countries, flooding of cable ducts during monsoon conditions can be a significant problem. These add to the cost of wired systems, making wireless a more attractive alternative.

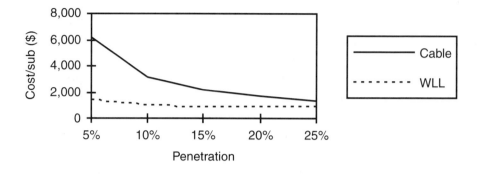

Figure 13.2   Relative costs of cable and WLL.

13.4   Current status of WLL

The world of WLL is currently experiencing dramatic growth; as an example, the number of WLL networks worldwide has risen from around 50 in 1996 to 130 in 1997 and analysts are predicting that there will be up to 200 million homes connected into the telephone system using WLL by the year 2000. Perhaps even more startling, the accountants Price Waterhouse predict that in the next 5 to 10 years, WLL will replace wired access as the predominant means of new connection.

The largest WLL network in 1997 was that of Spain, with around 800,000 subscribers. Most other WLL networks are either still in deployment or have just entered commercial operation. A good example of this is the Ionica network in the United Kingdom, which has been in commercial operation since summer 1996 and by spring 1997 had signed 15,000 subscribers, with rapid growth expected in coming months.

With a huge number of license applications and trials currently underway, it is certain that WLL subscriber figures are set to rise dramatically in the coming years. Some analysts' predictions for the number of WLL lines installed by the year 2000 are provided in Figure 13.3.

The predictions vary dramatically, showing the current lack of experience with WLL, but an average of the industry forecasts made in 1997 suggests that the market will be around 100 million homes connected, or "lines" by the year 2000, representing a market of $100 billion over the three year period. Beyond 2000, even higher growth is expected.

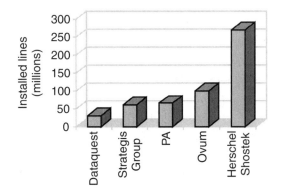

Figure 13.3 Analysts' predictions as to the number of WLL lines by the year 2000.

13.5 The differences between WLL and cellular

Put simply, WLL networks are simplified cellular networks. They use the same basic architecture and the same principles of radio transmission. However, since the subscribers do not move, they do not need location registers, means of handing over subscribers to different cells, means of allowing subscribers to roam to different networks, and strong means for combating fraud, for example.

However, WLL networks do have to provide a better quality of voice than cellular networks since they are designed to replace wired networks. They need to support data transmission at the rates that can be achieved through wired networks, they need to be extremely reliable, and they need to offer communications for a lower cost than mobile radio. These requirements result in different design decisions between cellular and WLL in terms of the way the radio channels are divided and the type of antennas used, for example.

13.6 WLL propagation

WLL propagation is mostly the same as mobile radio propagation, as described in Chapter 8. The differences generally result in the propagation being simpler and less problematic than for mobile radio.

WLL systems are generally designed to have a LOS path, which is literally just what it says; it means that if you put your head where the transmitting antenna is, then you can see the receiver. In mobile radio, a LOS between the mobile and the base station is a rare occurrence. However, WLL systems typically need to be designed such that a LOS is more frequently achieved because, at the frequencies used for WLL radio, waves do not bend ("diffract") well around obstacles. Hence, any obstacle tends to more-or-less block the signal, resulting in inadequate signal strength for good reception. Furthermore, WLL systems need to provide a voice quality comparable with fixed systems, unlike cellular, which is able to provide an inferior voice quality. In order to obtain a high quality, a strong signal is required that is typically only achieved through a LOS path.

This is not too problematic for WLL because the receiver units can be placed on roofs where a LOS is much more likely than would be the case for mobile radio. Establishing whether there is a LOS channel is straightforward—if the location of the subscriber antenna is visible (by eye) from the transmitter site, then there is a LOS. (In practice, visual surveys of potential transmitter sites are time consuming and computer models can help predict LOS paths.)

Despite the comments earlier concerning the need for a LOS channel, reflections can often provide an adequate signal level. An example of a reflected path is shown in Figure 13.4.

Reflection is caused when a wave strikes an object and is reflected back from it. Different materials reflect to a different extent, the amount of absorption by the material being a key parameter in determining whether a sufficient signal strength will be achieved. In practice, concrete and glass buildings provide quite good reflections. The receiver needs to be in just the right place in order to see a reflected image of the transmitter (compare with needing to be in the right place to see yourself in the mirror). However, many building surfaces are "rough" so that the reflection is scattered on hitting the building. This makes the zone where the reflections can be received larger but the signal strength weaker.

With mobile radio, the signal received is changing all the time, even if the receiver stays still, because moving objects around the receiver change the pattern of reflections, making the signal strength vary. Once a

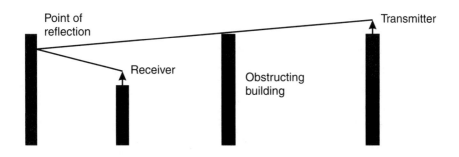

Figure 13.4   Example of a reflected path.

mobile starts moving, the signal strength changes very dramatically, as was seen in Figure 8.1.

With WLL there are fewer changes to the signal since most of the signal comes via a LOS path that will not be blocked by moving obstacles. However, the signal does vary and there are three key reasons why this occurs:

- Fast fading;

- New obstructions;

- Rainfall.

13.6.1   Fast fading

Fast fading was explained in Section 8.1. Broadly, if there is not a LOS path, or there are particularly strong reflections, then when large objects move in the vicinity of the receiver the fading pattern will change. If, say, the carrier frequency is a typical WLL frequency of 3 GHz (which means that the wavelength is 10 cm and hence the distance between fades 10 cm) and a bus which is momentarily reflecting a signal path into the receiver moves at 50 km/h (which is around 14 m/s), then 140 fades will occur every second, or one fade every 7 ms. In the worst case, the signal might decay momentarily to zero during a fade before being restored to full strength. There is little that the planning engineer can do to overcome this problem. Instead, the radio equipment designer must ensure that the equipment is able to correct the errors occurring from this phenomena.

13.6.2   New obstructions

New obstructions can occasionally arise. The most obvious example is the construction of a new building between the transmitter and the receiver, blocking a previous LOS path. Such blocking is often catastrophic in rendering the received signal too weak to be of use. Clearly, there is little that the radio planner or radio design engineer is able to do about this problem. The extent of the problem will vary depending on the rate of build in the area. The WLL operator should be aware of new buildings under construction and model their effect in order to look for possible solutions, such as using reflections, building an additional cell or repeater, or aligning the receiver toward the base station in an adjacent cell.

13.6.3   Rainfall

Rainfall can be a problem. Radio signals are attenuated by moisture in the atmosphere. The level of attenuation varies with the carrier frequency, the quantity of rain falling, and the distance of the transmitter from the receiver. The variation of attenuation with frequency is particularly strong and increases very dramatically at higher frequencies. At 3 GHz, the highest attenuation is around 0.06 dB/km; for a typical WLL path of say 6 km, the attenuation is only 0.36 dB (0.36 dB is equivalent to a loss of only 8% in the signal strength). However, at 10 GHz, the attenuation could, in extreme cases, be as high as 5 dB/km, or up to 30 dB over a typical link. Such a loss of signal would almost certainly make the link unusable. Hence, above around 5 GHz, in countries where heavy rainfall can be experienced, rain fading is a potential problem.

Broadly, the heavier the rain, the weaker the received signal. Hence, countries experiencing extremely heavy rainfall from time to time, such as those with monsoon climates, will be most affected by this problem. Rain fade results in an increased reduction in signal strength per meter; hence the further away the receiver, the greater the problem. Simply, those receivers on the edge of cells will be first to suffer during heavy rain.

Where rainfall attenuation is problematic, it must be allowed for in planning the system. Based on the frequency, the expected distribution of the rainfall, and the outage time that can be tolerated (if any), the additional path loss that needs to be accommodated can be calculated and the resulting reduction in propagation range determined. Cells then need to be placed closer together in order to take this into account. This, of course, increases the cost of the system.

Compared to mobile radio, WLL systems, even those with high data rates, tend to avoid ISI. This is because long reflected paths typically result from signals being reflected behind the mobile back into the mobile. The directional antennas typically used in WLL mean that such reflected paths are highly attenuated and thus fall below the receiver sensitivity threshold.

Problem 13.1

What are the three basic components of a WLL network?

Problem 13.2

What are the key differences between WLL and cellular?

Problem 13.3

Which technologies can provide higher data rates than WLL?

Problem 13.4

Why are WLL networks less expensive than copper networks to build?

Problem 13.5

How do video distribution systems differ from WLL systems?

14

WLL Technologies

14.1 Introduction

You have already learned about most of the WLL technologies. Manufacturers are currently offering WLL variants of GSM, CDMAone, DECT, PHS, as well as some proprietary technologies. When making a technology in a WLL variant, very little actually changes. The only real difference is that the base stations are designed to connect back to the PSTN rather than to a special switch for the mobile system and that there is no need for HLRs or VLRs, for example. The key issue often remains spectrum efficiency, but with WLL, the need for high voice quality and high reliability becomes more pressing than for mobile. Also, most users expect a higher data capability (e.g., at least 33.6 Kbits/s) from their fixed line, and hence WLL systems should be designed to achieve this.

As well as taking existing technologies and modifying them, some manufacturers have designed specific technologies for WLL. These tend to have better voice quality and a wider range of services than the modified cellular and cordless technologies. The proprietary technologies are all somewhat similar and it is not useful to look at them all in detail. Instead, one representative system, the AirLoop system, is described. Once this system is understood, the key parameters of the other systems can be rapidly assimilated.

14.2   The Lucent AirLoop WLL system

The Lucent AirLoop technology is a CDMA-based system developed for a wide range of users including residential customers and small and medium businesses. It is not suitable for large businesses who typically require tens of telephone lines into their building. It operates mainly in the 3.4-GHz band. It can provide a range of services to each user, from low bit rate speech through higher bit rate speech to full ISDN capability.

▼   ISDN

If you have ever paid any attention to television or mail advertisements from your local telephone operator, you will probably have noticed at some time that for perhaps twice as much as you pay for a normal line you could have something they call an ISDN line. It is typically marketed as offering much faster data transfer than you can achieve at the moment. So what is this high-speed telephone line and why do you need to know about it?

ISDN is actually not a telephone line, it is a way of organizing voice and data traffic in a fairly flexible manner. Data organized according to the ISDN format can be sent over any type of bearer—wire or radio—as long as the bearer has enough bandwidth to allow the signal through. ISDN is a set of rules, a little like those used to construct books. Everyone knows that the contents list comes at the front, the index at the back, and the chapters in the middle. Similarly, with ISDN, whichever unit receives the ISDN data knows where to find signaling information, speech information, and data in the ensuing burst of data that follows.

You need to know about ISDN in order to be able to decide what type of telephone line to buy from your operator or what capabilities to specify on your WLL system.

One of the strengths of ISDN is that it can provide a range of data rates depending on how much data users want to send and what the capabilities of the channel are over which that the data is going to be sent. The lowest rate ISDN channel is 64 Kbits/s, with a typical ISDN deployment providing a so-called 2B + D arrangement (this is known as basic rate ISDN access—abbreviated BRA). This includes two basic (B) 64-Kbits/s channels and one data (D) channel of 16 Kbits/s. Primary rate ISDN offers 30B + 2D channels, a total of nearly 2 Mbits/s, which is often used by large businesses but is not practical for WLL systems.

▲

The AirLoop system gets its flexibility from the way that channels are assigned to users. Each carrier provides 115 channels, each of which has a data capability of 16 Kbits/s. A user can then be given the relevant number of channels to make up the service that they require. For example, to support 32-Kbits/s ADPCM, two channels are used simultaneously (so every even bit is sent down one channel and every odd bit down the other channel and the receiver reassembles the data stream). For basic rate ISDN, nine channels are required simultaneously (four for each B channel and one for the D channel).

The top level network architecture is shown in Figure 14.1, and each of the main functional blocks of the network are described next. The basic architecture of the system is in line with the WLL architecture discussed in Chapter 13. There is a switch, base stations are linked to the switch and subscribers have radios that transmit signals to the base stations. However, there is a slight difference in that the base station function is partly divided in the AirLoop system. As you will see the base station has been split into two units called the *Central Access and Transcoding Unit* (CATU), and the *Central Transceiver Unit* (CTRU) (Lucent should be given some sort of award for selecting the most confusing and inappropriate names for the parts of their network). While the manufacturers have not made all their design reasons clear, at least part of the reason for this split concerns the manner in which base stations connect to PSTN switches.

To understand PSTN switches it is necessary to think about ordinary copper wire telephone networks. In such a network, each subscriber has a copper wire dedicated to them going from their house right to the local

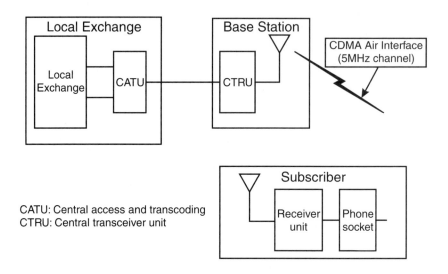

Figure 14.1 Radio access architecture.

switch. When the local switch has an incoming call for them it simply directs it down the appropriate telephone line. In order to interface with such a switch, a WLL system needs to provide a line for each subscriber. However, on the air interface, there is only a line for each on-going call with these lines being given to the subscribers as and when they wish to make a call. Somewhere in the WLL system, an interface unit needs to make these air interface channels look like dedicated copper wires. Such an interface is not too difficult to engineer, but the bandwidth requirements on each side of the interface are quite different. On the switch side, 64 Kbits/s is needed per user because the switches use PCM, a 64 Kbits/s voice coding system, as their default voice coding scheme. On the air interface side, approximately 32 Kbits/s times the Erlangs per user is needed. For example, if the user only generates 0.1E, then only approximately 3.2 Kbits/s per user is required. Hence, there is a difference in bandwidth requirements of a factor of 20 either side of this interface.

Transmission is expensive, even over copper cable, so it makes sense to place this interface unit right alongside the switch. Then the connection from the interface unit to the base stations can be made using much lower rate connections that cost less money, as detailed in Section 3.6. In the AirLoop system, the CATU forms the interface unit, which is placed alongside the switch, and the CTRU forms the base station.

Switch interfaces are due to change slowly over the coming years. Current interfaces conform to a standard known as V5.1. There is a new standard, which is being developed at the moment, known as V5.2, which will remove the need for this interface unit. Whether manufacturers will redesign their WLL systems to take advantages of this remains to be seen.

The functions of each of these boxes is now described in more detail.

14.2.1 Local exchange (termed a switch in GSM)

This contains the digital switching and network routing facilities required to connect the radio network to the PSTN.

14.2.2 CATU (termed a base station controller in GSM)

The CATU is colocated with the switch and performs the following functions:

- Controls the allocation of radio resources and ensures that the correct number of 16-Kbits/s bearers are allocated according to the service being provided, for example, four for 64-Kbits/s data transmission and two for 32-Kbits/s speech;

- For speech services, provision of transcoding (i.e., the changing between different speech coding formats) between various speech coding rates and the switched 64-Kbits/s PCM.

So basically, from the switch comes a 64-Kbits/s voice channel for each particular line. At the CATU, this is turned into 32-Kbits/s or 16-Kbits/s speech depending on what the operator has decided to use. This is then assigned the right number of 16-Kbits/s channels on the air interface and the signal sent to the base stations, either via leased line or microwave link, as explained in Chapter 3, in a form that they can transmit directly.

14.2.3 CTRU (termed a base station in GSM)

A single CTRU provides a single CDMA radio channel. The main functions are to transmits a radio signal to the subscriber unit and allow any signaling information, such as when the user presses a key on their telephone, to be passed transparently through to the switch.

So basically the CTRU acts as a radio transmitter, sending whatever information is received from the CATU to the subscriber using radio transmission. The CTRU does not do any processing on the signal, simply sending on whatever it receives. This makes sure that any signaling information that might be sent over the channel is not corrupted by the CTRU. This signaling information is what is used to provide additional services such as when you request the number of the last caller from the network by punching a certain sequence of numbers into the phone.

14.2.4 The subscriber unit (mobile in GSM)

The subscriber unit connects the subscribers to the radio network and comprises a radio receiver unit that is mounted on an external wall and an indoor unit providing a socket into which telephones can be connected. The two need to be connected together using copper wire.

14.3 DECT as a WLL technology

The system architecture of a DECT WLL system is shown in Figure 14.2. The main components of the system are the BSCs, the BSs, and the subscriber units. The BSCs are normally in a different location from the base stations, and hence there are TRXs to communicate between them. These transmitters either send their signal down a leased cable or using a microwave radio link; both types are shown on the figure, with the radio path at the top. The BSC is at the heart of the system; it controls calls originating or terminating in the radio network and provides connection to the local PSTN exchange. One BSC can handle 60 simultaneous calls and up to 600 subscribers.

In a typical deployment a number of BSCs are required. Typically, they will be located at the local exchange site because the link between the base station and the BSCs sends voice using 32-Kbits/s coding, while

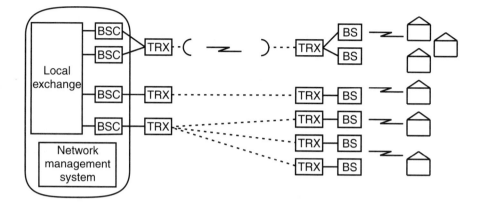

Figure 14.2 Architecture of a DECT WLL system.

the link from the BSCs into the network uses 64-Kbits/s coding. Hence, the transmission resources required are reduced if the base stations are deployed in this fashion.

The base station consists of a small cabinet and two antennas, spaced around 2m apart to achieve sufficient spatial diversity. Typically, omni-directional antennas are used, although directional antennas may be required where high range is needed. Each cabinet contains six DECT transceivers and a control unit that oversees the transceivers and multiplexes the call traffic on the connection back to the base station controller via the wire, fiber, or microwave radio transmission resource.

The unit installed at the subscriber's premises is composed of a DECT transceiver, a line interface unit, an antenna, a power supply, and a battery backup. The line interface allows an ordinary analog telephone to be connected into the unit. Multiple line units (up to 12) can be easily provided by connecting more line interfaces into the DECT transceiver.

DECT also has the possibility of deploying repeater units. These may be useful if mobility is needed in a limited area such as an office or shopping mall. The repeater unit works by receiving the data in one time slot during the receive half of the TDD frame and retransmitting it using a different antenna during the transmit half of the frame. It is essentially a combination of a DECT terminal and a DECT base station. Repeaters are deployed in two ways.

1. The coverage enhancement unit uses multiple directional antennas to extend the range of the base station, providing coverage in shadowed areas and connecting to the subscriber units in the normal manner. It can also be used to overcome the problem of new buildings appearing after the network was planned.

2. The residential fixed repeater unit uses a directional antenna to communicate with the base station but then reradiates via an omni-directional antenna creating a local microcell.

Although repeaters provide some flexibility, they have two significant faults. The first is that they add delay equivalent to the time delay between the receive and transmit parts of the TDD frame, some 10 ms in total, which is slightly disturbing when speaking, especially if there is already some other delay on the phone line (e.g., resulting from a satellite link being used). The second is that they are inefficient of resource, requiring twice as much spectrum to carry the call as if a repeater was not used. Nevertheless, in some situations they can be beneficial.

For a number of years after the standard was introduced, DECT failed to take off. However, in recent years penetration has increased substantially, particularly in Germany. It may be that with success in the WLL market DECT will sell substantially around the world.

As mentioned in Section 12.2, ISI can limit the range of DECT systems to only around 300m. However, in the case of WLL, where directional antennas are used at the subscriber equipment, the range can be substantially extended. Figure 14.3 shows that, with a directional antenna with an overall beamwidth of 20 degrees, the subscriber unit can be positioned much further from the base station before ISI will become problematic (simple mathematics shows that the distance to the subscriber unit can be increased to 23 km in this case). DECT has a relatively low transmit power, and hence it is more likely that the range in a WLL deployment would be constrained by inadequate signal strength rather than ISI.

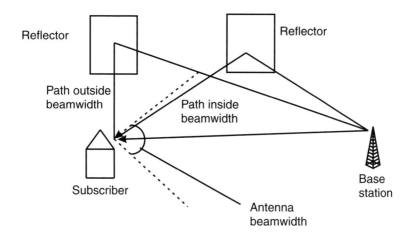

Figure 14.3 Use of the antenna beamwidth to discriminate against long reflected paths.

Problem 14.1

What is the key difference between, say, a GSM system designed for mobile use and one designed for WLL use?

Problem 14.2

Is ISDN a format for cables or radio transmission?

Problem 14.3

Why is the provision of a permanent line to each subscriber, as in a wired network, not used in a wireless network?

15

The Future of WLL

15.1 Introduction

This chapter is has been divided into four areas. The first looks at possible technical advances in WLL technology, enabling better and lower cost offerings. The second looks at the capabilities of the other access technologies that are in competition with WLL. The third section considers how the users' demands might change as the much heralded Information Society arrives. The final section attempts to draw some conclusions from all these predictions as to how future demand for WLL systems might develop.

15.2   Technical advances in WLL

Consumers are accustomed to computers that double in power and capacity almost annually and a seemingly invincible technical revolution. Broadly, though, such revolutions have not occurred in the world of wireless. Although there has been a transition from analog to digital mobile telephony, the consumer will not have gained that much (perhaps a slightly better voice quality and better data transmission) and this transition will take place over a 15-year period. Plans for third-generation mobile radio systems, discussed in Chapter 17, show little more than an ability for the phone to master multiple different standards. Equally, early predictions of the spectrum efficiency of the newer systems have generally proved to be incorrect, with GSM only providing marginal capacity increases over analog cellular systems.

Hence, significant technological advances enabling much more efficient use of the radio spectrum and providing valuable additional benefits to users seem unlikely. WLL systems will add additional features and gain flexibility, but broadly no major breakthroughs should be expected.

By far the most important development will be the achievement of large economies of scale through worldwide penetration that will lead to lower prices further increasing the cost advantage of WLL systems. At present, subscriber units cost perhaps $1,000 for the proprietary systems whereas GSM mobile phones cost around $200 each. In principle, there is little reason why WLL subscriber unit prices should not fall to the level of mobile phones, or near those levels, as long as sufficient economies of scale can be achieved. This is a reinforcing trend; as prices fall, demand increases, resulting in greater economies of scale and further reductions until the equipment cost falls to the minimum manufacturing cost. WLL will undoubtedly follow this path since it has already taken the first steps along it, and orders are increasing even at current prices. However, it will not follow it so quickly as the mobile systems due to the lack of standardization and resulting lack of competition. Nevertheless, within a decade, subscriber units should cost less than around $400 each in real terms (and the subscribers will typically not pay this—the unit will be subsidized by the operator in the same manner that mobile phones are subsidized in most countries).

15.3 Technical advances in other access techniques

The key question mark facing WLL systems is whether enhanced means of sending large amounts of information down existing copper wires will develop to such a level that subscribers will be able to receive video on demand and similar material through their existing telephone line in a cost-effective manner. Such a development would allow the wired operators to subsidize their cost of telecommunications with additional services, making market entry difficult for WLL operators in areas already well served with fixed telecommunications. The success of such approaches depends almost entirely on the quality of the existing lines, which varies from region to region and country to country. Prediction in this area is almost impossible; however, it seems likely that the cost of high bandwidth provision over copper will not fall to the level where it becomes attractive for mass-subscription use such as video on demand. Instead, it will be of most interest to a limited number of people working from home and willing to pay perhaps as much as $1,000 in total for connection. This limited market will not achieve the economies of scale necessary to drive the prices down, so high-bandwidth copper systems will remain a niche application, not seriously threatening WLL operators.

Internet telephony is currently an application that is concerning many existing operators. The concept is to use advanced speech coders to compress and segment the voice and then send the resulting speech packets over the Internet. The advent of this technology is entirely artificial due to the fact that all Internet calls are charged at local rates despite the fact that many result in international flows of data. This, in fact, reflects the true cost of telephony provision, where the local loop section represents the vast majority of the cost and the national and international trunk contribution is almost insignificant. However, fixed operators persist in inflating international prices in order to increase profitability and in some cases subsidize the local loop. If Internet telephony works (and it is unlikely to for long, even if it does work at first, as the network simply becomes swamped by voice as well as the current heavy data load) the net result will simply be that fixed operators will re-balance their tariffs so that Internet telephony is no longer worthwhile. The only advantage then will be the ability to link voice with exploring a site so that a user could speak to the owner of the site they are surfing if required.

Even that discussion is irrelevant for WLL. Whether the voice service is through the voice network or across the Internet, it will still be carried along the same WLL lines to the network and can still be tariffed at the same rate. Hence, Internet telephony holds little threat for the WLL operator.

15.4   Changing user demand

It is often said that the world is becoming an Information Society, with huge amounts of information being made available to individuals and passed rapidly around the world. The Internet is held up as the first manifestation of such a development. If true, users will increasingly demand bandwidths of megabytes per second as services are provided that require them. Certainly, the Internet is increasingly requiring high-bandwidth access, and as file sizes and hard disks get increasingly large, it is undeniable that user demand for data is increasing.

However, it is easy to get carried away with this approach. Most people rely on broadcast entertainment, which will increasingly provide more numerous channels and accompanying data. This will be delivered via the broadcast and not the telephony channels. Generally, those who require large files will only require them periodically (because they will need time to digest what they have received). Finally, and most importantly, users will only want such services if they can be provided at a low price; for example, most Internet users could benefit from an ISDN connection but do not choose to have one because the additional cost exceeds the value that an ISDN line will add for them.

Therein lies the problem. Ubiquitous provision of high-bandwidth services will require enormous upgrades to the world's telecommunications network, particularly switches and packet servers and certain parts of the interconnecting network, not to mention the access network. Such upgrades cost money and as yet there are few users who value the Information Society enough to pay the additional cost required. Within the next 10 to 20 years there may be many users who would like a high-speed connection but few who are prepared to pay for it. The user demand will be for information at a reasonable cost that will encourage information providers to compress data and to use graphics more sparingly than otherwise.

Indeed, perhaps bizarrely, at the moment the user is actually prepared to pay less for higher bandwidth services. Figure 15.1 shows approximate current levels paid per minute for a range of different services by those in developed countries.

As can be seen in Figure 15.1, the highest value services are typically also those that can be achieved by existing narrowband systems. With the present set of services, users are only paying something like 1/1000th of the amount per channel for broadband services when compared to narrowband services. Since the cost of provision goes up in line with the amount of bandwidth (since the bandwidth could be used for multiple narrowband services rather than one wideband service), there is presently a massive gulf between cost of provision and expectation of cost. This will slow the widespread adoption of broadband services for some years to come.

Although the demand for high-bandwidth provision may not rise dramatically, the use of the access network for applications other than voice will change significantly. Internet access, including applications such as

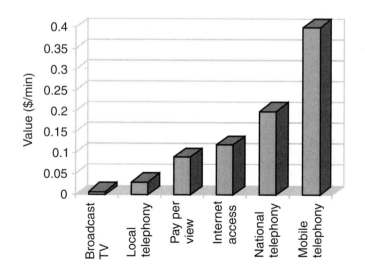

Figure 15.1 Variation in value for different services.

home shopping, will dramatically increase the traffic generated by a home in that the phone line will now be used for a much greater percentage of the day. Modems will become a standard fit within computers and probably also within televisions. This will fuel the demand for two or more lines per house and will also increase the revenue received per line by the operator.

One final area worthy of mention is the increased environmental awareness that will make the laying of additional cables increasingly problematic. Already, cable operators are criticized for damaging tree roots. Once they are forced to pay compensation for the inconvenience that their digging causes, wireless will become the preferred option for most connections.

15.5   Predictions for WLL

At the start of 1997, a number of analysts such as Ovum and PA made projections of the WLL marketplace that were shown in Figure 13.3. These suggested that by the year 2000 over 10% of all lines being installed will be wireless, while others have predicted that the number of wireless lines installed per year could overtake wired lines before the year 2005.

Finding the reasons for such optimistic predictions is not difficult. The key factors are:

- The cost per line for WLL is currently lower than for wired systems and will fall further with economies of scale;

- The demand for telecommunications around the world will increase significantly in coming years;

- Fragmentation and the development of market niches in the telecommunications market will allow a range of different operators to coexist.

The predictions do not hinge on any technical or commercial breakthrough, although they do require that other access technologies do not develop unexpectedly.

Of course, there will be some problems. Some operators will fail in the first few years through a misunderstanding of the market or through the deployment of inappropriate equipment. Problems will soon emerge with lack of radio spectrum and this will, effectively, increase the cost of radio spectrum, either directly, through auctions, or indirectly through the impossibility of obtaining more, forcing greater network costs as smaller cells are deployed. This will be partially solved as higher frequencies are used; but as with mobile radio, shortages of spectrum will remain a permanent problem, limiting subscriber numbers and the data rates that can be offered.

Nonetheless, the future of WLL is certain to be a bright one.

16

Satellite Systems

16.1 Why use satellites?

Only a very small percentage of the world is covered by cellular systems. In most developed countries only the rural areas are without cellular coverage, but many developing countries have very limited coverage because the population is not able to afford cellular calls. Even where there is coverage, it could be using a number of different standards. This means that when traveling to a different country, the cellular coverage may be useless unless you carry a wide range of mobiles; for example, the global executive who travels to London, New York, and Tokyo would need three different phones: a GSM phone, an AMPS phone, and a PDC phone, respectively.

The story goes that a senior cellular executive and his wife went on holiday to some remote island. His wife needed to make a phone call but found that her cellular phone did not work. "Why can't you design a phone that works everywhere?" she asked her husband, and hence the satellite phone concept was formed.

Satellites are the only way to cover the whole world. Cellular systems will never achieve global coverage because each cell can only be a few kilometers in radius, making the cost of covering large areas completely uneconomic. Satellite cells can be many hundreds of kilometers in radius, and hence the whole world can be covered with a relatively small number of cells.

16.2   How satellite systems work

Satellite systems, at the simplest level, work just like cellular systems except that the transmitters are now up in space. Paging and location updating, for example, are all just the same as for cellular. The biggest difference is caused by the fact that the transmitter is so far away.

Early satellite systems and satellite TV systems use satellites in what is known as geostationary orbits. This is an orbit where the satellite goes round the Earth at the same speed as the Earth rotates. So when looking at the satellite from the Earth it appears to stay in the same place. This is highly desirable for satellite systems where dishes are used to receive the signal because it means that the dish can stay pointing in the same direction throughout the lifetime of the service. However, for personal communications geostationary orbits have three key problems.

- Delay: To be in a geostationary orbit, the satellite needs to be 36,000 km above the Earth's surface. To go up to a satellite and come back down again takes a radio signal around one-fourth of a second, which is quite a noticeable and annoying delay when trying to have a conversation with someone.

- Power: To transmit a signal 36,000 km requires a lot of power, and this rapidly drains the batteries of mobile phones.

- Capacity: From a geostationary orbit, most of the world can be covered with three or four transmitters. Remember back to Chapter 3 where using a number of cells was one of the key ways to increase capacity. With very few cells, the capacity of a system using geostationary orbits is very low.

A solution to this problem is to use different orbits. Some of the satellite systems proposed are using what are termed *low earth orbits* (LEOs). These are orbits only some 500- to 1000-km above the Earth's surface. Using such low orbits dramatically reduces the delay and the power consumption to manageable levels. These lower orbits also have smaller cell sizes, and hence the total system capacity can be higher. A typical LEO system might have around 70 satellites for global coverage, providing over 20 times the capacity of a geostationary system (but at a much greater cost because of the much greater number of satellites required).

LEOs have one big problem; the satellites are not stationary in space. In fact, in a LEO, they move overhead extremely quickly; in some cases satellites can take as little as 6 minutes to pass overhead. This means that directional mobile antennas cannot be used since pointing them at the satellite would be extremely difficult. This is not too serious a problem when in a low orbit because the gain provided by antenna directionality is not needed when the satellite is much closer than a geostationary orbit. More problematic is that numerous satellites might pass overhead during the course of a call. Handover is therefore required between different satellites. In practice, this handover is simpler than in a cellular network where the network has to deduce to which cell the mobile should be handed over. In the satellite system, the system knows which is the next satellite coming along to cover that area, so it knows exactly to where the calls need to be handed.

16.3   Will satellite systems be successful?

Satellite systems have one major advantage—with a satellite phone you can make a call from anywhere in the world, regardless of what the local cellular infrastructure is like. They have two major disadvantages—first, it costs a lot more to deploy a transmitter in space than on the ground and, second, the capacity of a satellite system is only a tiny fraction of that of a cellular system. (You also cannot use satellite systems indoors which will be a problem for many users.)

To understand the capacity in more detail, remember that the system capacity is proportional to the number of cells in the system. Although each satellite uses advanced antennas capable of forming a number of relatively small cells on the surface of the Earth, each satellite can only handle around 3,800 simultaneous calls. With around 70 satellites, this means 260,000 on-going calls throughout the world, and at cellular traffic levels of around 0.05E per subscriber (i.e., the average subscriber uses the phone 5% of the time during the busy hour), around 5.3 million subscribers is the maximum system capacity. But as was shown in Chapter 3, this is about the capacity of a single cellular network in a single country. In 1997 there were over 300 cellular networks in the world, and that number will probably be between 500 and 1,000 by the time the satellite systems are in full operation. So the satellite systems have only around 1/500th or 0.2% of the capacity of cellular networks. Clearly, satellite systems are not going to steal significant traffic from cellular systems.

A satellite system with around 70 LEO satellites is predicted to cost around $4 billion to build and operate. This is perhaps twice as much as a cellular system with equivalent capacity, although system costs are highly variable. In very simple terms, the success of satellite systems depend on there being some 5 million people around the world who are prepared to pay at least twice the price of cellular calls for the benefit of being able to make a call anywhere in the world. For these people, it is necessary to look mostly toward the United States. The United States has different cellular standards from most other countries and so it is more difficult for U.S. subscribers to roam around the world than it is for, say, European subscribers, where they can take their GSM terminals to over 100 other countries. Coupled with users on ships and aircraft, in remote areas of the world, for example, finding these 5 million users does not look overly problematic.

A further development that will increase the attraction of satellite systems is the dual-mode cellular-satellite phones that are being developed. These are phones that will work on a cellular network whenever there is one available. When there is not, they will switch to satellite mode. Clearly, this brings benefit to the user—their cost of communications in urban areas remains the same and they are now able to stay in touch in rural areas as well. The operator may also gain some benefits in that they can offer their subscribers excellent coverage without the cost of building out their network in rural regions. Indeed, some cellular operators have taken shareholdings in some satellite operators on the basis that future networks will be formed from combined cellular and satellite systems.

At the moment there is a range of satellite systems proposed by different operators. Perhaps the most well-known is Iridium, for which one of the major backers is Motorola. Iridium will use 66 satellites[1] in a LEO to provide a worldwide service. Around half the satellites were launched during 1997, with full service expected toward the end of 1998; although some commentators expect service to take a little longer than this to launch. Other operators are working on different variants using fewer satellites in slightly higher orbits to provide a cheaper but lower capacity service. It remains to be seen which will be successful and whether there is a sufficiently large market for numerous satellite systems in the world.

Problem 16.1

What are the advantages of geostationary orbits?

Problem 16.2

Why are geostationary orbits not suitable for mobile satellite systems?

Problem 16.3

How high above the Earth is a LEO?

Problem 16.4

What would a dual-mode GSM/Iridium phone do?

---

1. Iridium was originally proposed with 77 satellites. The name was selected because Iridium is the atomic element with 77 electrons circulating around it. Subsequently, for cost reasons, the number of satellites was reduced to 66, however, the name Iridium was retained. Some think that this is because the atomic element with 66 electrons, dysprosium, sounds more like a stomach upset rather than a global communications system!

17

The Future of Cellular

17.1 What could be better than current systems?

It is, of course, true that those who are most able to accurately predict the future invest accordingly and make their fortunes, while those less able resort to attempting to make money from telling others what their vision of the future is. Further, most authors wisely preface predictions with some of the many examples of amusingly inaccurate predictions that have ever been made. All these warnings are relevant here. Nevertheless, predictions are interesting to read and, if treated with caution, can provide useful stimulus to thought.

To most cellular users who are just moving from analog cellular to GSM, the digital system appears to be a marvelous advance. Voice quality is better, data and faxes can be sent over the phone, users can roam to most countries in the world, and the phones are getting smaller and more feature-packed all the time.

However, there are a number of flaws with the current systems. One of the key flaws is the inability to roam across the entire world with one single mobile. Another is the limited data rates that can be achieved with digital cellular systems. The final problem, which is not immediately apparent to many users, is that the digital cellular systems do not meet the needs of many professional users of radio systems such as police and railways.

Cellular systems are generally thought to have come in generations. Analog cellular is said to be the first generation. Digital cellular systems, such as GSM and CDMAone, are said to be second generation. The next systems, which are currently being planned and will aim to solve the problems listed previously, will be termed third generation.

17.2   The third-generation ideal

First, before starting to explain the third-generation ideal, it is important to understand the terminology used. Within Europe the next generation has been termed *Universal Mobile Telecommunications System* (UMTS), while at a worldwide level it has been termed the *International Mobile Telecommunication System for the Year 2000* (IMT2000).

The vision of UMTS that is shared to a large extent by those outside Europe has been articulated in many, slightly differing ways but can be summarized as "Communication to everyone, everywhere." Communication in this instance might also include the provision of information. It will be a system that everyone will use regardless of whether they are in the office, at home, or on a airplane. To achieve this goal the system must essentially provide an extensive feature set to appeal to all the current disparate users of different types of radio system:

- Cellular users who require high voice quality and good coverage;

- Users who currently deploy their own private systems (often known as PMR users) and require group and broadcast calls where all users can hear what one person is saying;

- Paging users who require a small terminal and good coverage;

- Cordless users who require excellent communications with high data rates when in the office or home;

- Satellite users who require truly worldwide coverage;

- Users in airplanes who currently have limited telephone availability (especially to receive calls);

- Data users ranging from telemetry to remote computer network access

and ubiquitous coverage with a wide range of cells such as:

- Satellite cells covering whole countries;

- Macrocells covering a radius of up to 30 km;

- Minicells covering up to around 3 km;

- Microcells covering a few streets;

- Picocells covering an office, a train, or an airplane, for example.

Major players in this area (chiefly manufacturers and cellular system operators) are postulating that this will be achieved through a system with the following characteristics:

- An "adaptive air interface" such that access methods such as CDMA, TDMA, and FDMA can be selected by the mobile phone as appropriate with bandwidths dependent on the service required and a data rate of up to 2 Mbits/s available in some locations;

- Mobile phones that can have their "operating system" downloaded to allow for network evolution;

- An architecture based on intelligent network principles.

▼ When is a network intelligent?

In recent years, much work has gone into developing intelligent networks. Unfortunately, the use of the word "intelligent" is mostly inappropriate, so an intelligent network might not be what you think it is.

Broadly, an intelligent network is one that easily allows new features and services to be added with minimum disruption. An intelligent network contains a toolkit of features that the service designer can call upon. Items in the toolkit include:

- The ability to recognize when a user has hit a particular number on their phone;

- The capability to make a spoken announcement when the words to be spoken are provided as text;

- The capability to forward and redirect calls in a flexible manner.

As an example, imagine that as an operator you decided that your subscribers really needed call forwarding according to time of day so that if their number was called and they did not answer it, the call would be redirected to their mailbox outside office hours and to their office phone in office hours. You would need to instruct the switch to ask the users for the hours during which they would like the calls to be forwarded to different places by playing an announcement of the form "please enter the time that calls are to start to be directed to the office using the keypad on the phone, press the buttons corresponding to the hour you want redirection to start." The switch would then wait for the user to press the appropriate keys and then would continue to ask them for other parameters.

This capability requires a particular network structure able to store software and activate toolkit elements accordingly. Such a design has been called an intelligent network. Actually, intelligence in networks has fallen ever since the start of the telephone system toward the end of the nineteenth century. At that time, there were switchboard operators. When someone called the switchboard, they would ask to be connected to "John," not a particular number. The operator might tell them, "Oh, John is visiting his mother today, I'll put you through to her." That was an intelligent network.

▲

The system must also integrate seamlessly with the fixed network such that users receive nearly identical services whether they are using fixed or mobile phones. Current predictions as to the timescales of third-generation systems vary slightly, but typically it is assumed that standardization will be complete by 1999, the first product will be available by 2002, and the product will be widely available by 2005.

One of the prerequisites of the third-generation system was that it would be agreed-upon worldwide as a single global system, achieving the design aim of international roaming. However, recently this has started to look like an increasingly unlikely outcome. There are broadly three key players in third-generation work: Europe, the United States, and Japan.

- Europe would like to see GSM evolve to become the third-generation system and is proposing an advanced version of GSM with a wideband CDMA air interface.

- Japan has a different agenda. They have broadly failed to make any impression on the world scene with their first- and second-generation systems and want to make sure that they do not fail with the third generation by launching a new system as early as possible, based on something quite different from GSM and also using a wideband air interface.

- The U.S. authorities have the view that as little as possible should be standardized because standards prevent innovation and consumer choice and that the standards making is often not performed by the most appropriate bodies. Some U.S. manufacturers have gotten together and are promoting an evolution of CDMAone that may become another contender for third-generation systems.

The pressures from these different bodies have now risen to the extent that most observers expect third-generation systems to include a variety of standards and for third-generation phones to be multistandard so that they can work in whatever country they are used.

17.2.1 GSM evolution

GSM is currently undergoing considerable enhancement. Within the so-called GSM Phase 2+, which gathers all the additions to the GSM standard, there is a wide range of features, including:

- The *advanced speech call items* (ASCI) consisting of group and broadcast calls and priority and pre-emption;

- A packet data service known as *general packet reservation service* (GPRS);

- A higher data rate service known as *high-speed circuit-switched data* (HSCSD) capable of providing data rates of up to 64 Kbits/s through concatenation of a number of time slots;

- Intelligent network capabilities, known under the misleading title of *customized applications for mobile network enhanced logic* (CAMEL)—a typical use for this service is the support of operator specific services while roaming;

- The enhanced full-rate codec that aims to provide significantly enhanced voice quality using a single timeslot (as at present for voice communications);

- Combined DECT/GSM handsets capable of seamless roaming between the two technologies.

Outside the scope of the Phase 2+ standardization work there are other developments such as dual-mode satellite-GSM handsets. These have been proposed by S-PCN operators (e.g., Iridium), who realize that the market for satellite services will be limited unless handsets interwork with terrestrial infrastructure, where available.

Assuming all these Phase 2+ features come to fruition, GSM will provide most of the facilities required by most of the users of radio systems. Considering all the users for whom third-generation systems are supposed to provide a service it can be seen that:

- Cellular users are already catered for.

- Private users will be supported by the ASCI features.

- Paging users can already be supported by SMS.

- Cordless users will be supported through the DECT interworking.

- Satellite users will have dual-mode GSM satellite terminals.

- Users in airplanes may not be fully supported.

- Data users have a wide range of packet and circuit services up to 552 Kbits/s where DECT interworking is supported and where DECT coverage is provided (which will only be in high-density areas).

It could be argued that with all these features, satellite interworking and the near-global roaming capabilities of GSM, it will soon fulfill all the goals of UMTS.

Whatever the route, by the year 2005 expect there to be a new generation of cellular systems. These systems will have a much wider range of capabilities, will support higher data rates, and will be able to communicate with satellites when out of the range of cellular systems. However, roaming may still be difficult, and it may even be that there will be more third-generation standards than there are second-generation ones.

17.3 Mobile/fixed convergence

Many in the telecommunications industry have been talking for some time about fixed/mobile convergence. Perhaps "convergence" is a misnomer, "mobile dominance" might be a better way to describe the vision of those articulating such a future. Simply put, why not use your mobile at home as well as when roaming. As long as the quality is sufficient and the call charges are low, this makes establishing contact with an individual much simpler.

There are many flaws with this argument that are increasingly emerging with time. These include:

- The cost of providing sufficient coverage such that mobile phones can reliably be used indoors has typically proved to be uneconomic to the cellular operators.

- Cellular operators have been unable to reduce prices to the level of those in the fixed network (which is unsurprising since they need to pay the fixed network operator an interconnection charge on mobile calls), so cellular prices have remained higher than fixed ones.

- Cellular voice quality remains below that of fixed phones.

- Cellular data rates are lower than those that can be achieved over fixed phones.

- Cellular availability and reliability is still a long way below that of fixed phones.

- Users dislike not having a fixed phone because they are concerned that one member of the family might take the mobile phone, leaving the rest of the family without communications.

Although some of these will be solved with time, there are sufficient problems to ensure that fixed-mobile integration will not occur in any meaningful way in the foreseeable future. Instead, intelligent network platforms will perform call redirection between fixed and mobile phones depending on a wide range of options so that single numbers can be used and single bills provided. Indeed, such platforms are already here and are used for premium "personal numbers" in some countries.

17.4   The longer term future

Where mobile phone systems are likely to go in the longer term future can be predicted by asking what the users want from a mobile phone. Some want to be able to transmit video images; computer files; all sorts of information associated with controlling their house, or car, or whatever else they own, and medical information, for example. The list is almost endless. What is in this list is not really an issue for the mobile phone, it is just the size of it. The mobile phone of the future will need to be able to send very large amounts of data, in a flexible manner that allows it at one moment to be programming the video and at another performing video-conferencing. High-data capability and flexibility will dominate the design.

This book discussed many techniques for achieving higher capacity than offered by GSM. These included using CDMA as a multiple access method, frequency hopping, discontinuous transmission, and power control, for example. But none of these techniques promise really large gains in capacity—gains of a factor of 10 or 100, which will be required for these systems of the future. The only technique thatcan provide these sorts of gain is the use of very small cells, the microcells discussed in Section 8.5, where it was pointed out that near-infinite gains in capacity could be achieved if extremely small cells were deployed. However, small cells are very expensive, which is currently limiting their widescale introduction.

The system of the future will need to tackle this problem. Small cells (indeed, any size cell) are expensive because of three costs that add together in approximately equal portions over the life of the cell, namely:

- The cost of the equipment needed for the cell site itself;

- The cost of renting space on a building in order to install the cell;

- The cost of renting a leased telephone line to link the base station back to the switch.

To tackle these problems, a new design methodology is required. First, the base stations must be designed in a different way that lends itself to mass production. Such base stations might even be simpler than those of today; for example, frequency hopping could be dispensed with if there were so many small cells that sufficient capacity was achieved without it. By anticipating sales of millions of base stations rather than the thousands achieved today, manufacturers could design the base stations in such a manner that the economies of scale could drive their prices down to a fraction of those of today.

To reduce the cost of rentals, base stations need to be much smaller. The smallest ones today are about the size of a small filing cabinet. Base stations of tomorrow need to be miniaturized to something no larger than a shoe box. Given the reduction in size that has been achieved with mobiles, this should not be too difficult with base stations—there has just not been the incentive to do it to date. When they are this small, rentals will be cheap because they take up almost no space.

The solution to the cost of leasing a telephone line may come from elsewhere. The new technologies that allow large amounts of data to be sent down normal telephone lines should dramatically reduce the costs of linking these base stations back to the switch. These technologies, which were called ADSL, were briefly mentioned in Section 13.2.

With small base stations and low-cost line rental, mobile communications provision will become ubiquitous and high capacity, providing the facilities required for mobile radio in the future.

Bibliography

THERE ARE MANY excellent texts on cellular covering all aspects of cellular systems in more technical depth than was possible here. Nearly all the texts are more complex than this book and require a good understanding of the concepts in this book as well as, typically, a reasonable mathematical grounding.

An excellent starting point and essential reading for anyone working for cellular operators or government regulators is a text on the history of cellular radio and the lessons that can be drawn from the last 10 years. Superbly written and full of detailed information, the following book comes highly recommended:

Garrard, G., *Cellular Communications: Worldwide Market Development*, Norwood, MA: Artech House, 1997, ISBN 0-89006-923-9.

For those who want to take the next step in technical understanding of radio systems, an excellent book to move onto is:

241

Mouly, M., and M.-B. Pautet, *The GSM System for Mobile Communications*, published by the authors, 1992, ISBN 2-9507190-0-7.

Mouly obviously focuses on GSM and covers much of the material presented here, but in more detail, and in particular spends more time on the protocols and detailed design issues. As already explained, a good understanding of GSM means a good understanding of most cellular systems. The book is not mathematical, and readers who have understood this book should be able to readily understand Mouly and Pautet. An alternative to Mouly, which covers a similar range of topics and also comes recommended, is:

Redl, S, M. Weber, and M. Oliphant, *An Introduction to GSM*, Norwood, MA: Artech House, 1995, ISBN 0-89006-785-6.

These authors have also written *GSM and Personal Communications Handbook*, which was unavailable at the time of writing but promises to cover the services offered by GSM, the competitors to GSM, and the evolution of GSM. This will be published by Artech House during 1998 and is intended to be a sequel to their existing book.

Another interesting book, which is also nonmathematical and easily understood by those who have grasped this book, is:

Horrocks, R., and R. Scarr, *Future Trends in Telecommunications*, New York: John Wiley, 1993, ISBN 0-471-93724-X.

This covers a wide range of topics including advances in computer chips and memories, switching systems, and channel coding, for example. It will be of particular interest to readers who want a wider perspective than that offered here as, in addition to covering systems such as GSM and UMTS, it also provides information on cable and satellite systems, intelligent networks, virtual private networks, standards, regulation, numbering, and network management. As can only be expected in a book covering so much material, each area is only covered in brief but as a wide ranging reference it is excellent.

The next stage, beyond these nonmathematical and quite general books, is:

Webb, W., and L. Hanzo, *Modern Quadrature Amplitude Modulation*, New York: John Wiley, 1994, ISBN 0-7273-1701-6.

Although centered around modulation techniques, this book covers cellular and wired systems in some detail in the early parts of the book before covering more detailed areas such a propagation, modems, and equalisers. The book provides a good blend of text discussing the issues, coupled with mathematical formula where required. Naturally, since the author is the same, it is written in a similar style to this book, which may be a help to some readers.

A text that compliments this book quite well, in that it covers mostly the same topics but assuming the basic knowledge provided here, is:

Pahlavan, K., and A. Levesque, *Wireless Information Networks*, New York: John Wiley, 1995, ISBN 0-471-10607-0

The book is substantially more technical and does require a reasonable understanding of mathematics to be fully grasped. But for those who want to read further on a few selected topics such as radio propagation and spread spectrum, then it forms a worthwhile text. An alternative to Pahlavan, and one that has been a standard student text for some time, is:

Proakis, J., *Digital Communications*, New York: McGraw-Hill, 1983, ISBN 0-07-050927-1.

This, however, is highly mathematical, has little explanatory text, and should certainly only be attempted by those comfortable with calculus.

For those who want to know more about the network side of the radio system, especially how switches work, how optical fibers transmit signals, and how whole networks are put together and synchronized, the following book is worth reading:

Bellamy, J., *Digital Telephony*, second edition, New York: John Wiley, 1991, ISBN 0-471-62056-4.

Although this book does include a significant amount of mathematics, mostly the flow of the text can be understood without grasping the mathematics and the book is well written and structured.

Beyond this, the texts become increasingly difficult and increasingly specific to particular areas. For those who want to know more about Erlangs and blocking and queuing theory consult:

> Schwartz, M., *Telecommunications Networks*, Reading, MA: Addison-Wesley, 1987, ISBN 0-201-16423-X.

It forms a standard reference text, but typically this area will not be of great interest to most readers. Propagation at higher frequencies and to satellites is well covered in:

> Freeman, R., *Radio System Design for Telecommunications (1 to 100 GHz)*, New York: John Wiley, 1987, ISBN 0-471-81236-6.

But again this is not a major area of cellular and will be of limited interest to most.

Cordless communications is very well covered by a comprehensive textbook analyzing all the different cordless technologies and providing chapters explaining the different roles of cordless systems:

> Tuttlebee, W. (Ed.), *Cordless Telecommunications Worldwide*, New York: Springer-Verlag, 1997, ISBN 3-540-19970-5.

A book covering DECT is due to be published by Artech House during 1998 and will provide more information about this particular standard than Tuttlebee is able to do in his more general text.

Because of the relative newness of WLL, there are far fewer references. There is currently only one book that covers the topic in detail:

> Webb, W., *Introduction to Wireless Local Loop*, Norwood, MA: Artech House, 1998.

This covers WLL in detail, from the technologies used to a case study of a complete WLL network design. It is nonmathematical and would be easily assimilated by those who have mastered the material in this book, especially since the author is the same. It is certainly the most appropriate way forward for those wanting to know more about WLL.

The only other book that can be recommended in the area of WLL is:

Calhoun, G., *Wireless Access and the Local Telephone Network*, Norwood, MA: Artech House, 1992, ISBN 0-89006-394-X.

This book has the rather serious drawback of having been written before WLL started to gather pace and hence much of the discussion predicts the arrival of WLL rather than talks about actual systems. Nevertheless, as background reading to the competitive environment in the United States, to the legal issues associated with WLL, and to see how rapidly ideas change in this area, it could be of interest.

Model Answers

2.1   Cells are approximately circular because a transmitter radiates equal signal in all directions, which all things being equal will travel the same distance, resulting in a cell in the shape of a circle. In real life, obstructions such as hills or tall buildings tend to prevent radio signals from traveling the same distance in all directions, and in some cases directional antennas are used to direct the signal into a particular direction.

2.2   The three main components are the switching system, which ensures that calls reach the right subscriber; the base station system, which handles the radio transmission; and the mobile receiver with which users can communicate with the network.

2.3   The VLR prevents signaling relating to the position being sent internationally in the case that a mobile has roamed to a different country and by keeping information within the MSC area where the mobile is located allows call requests to be handled more rapidly.

2.4   A GMSC is the interface point to the PSTN and makes sure that calls reach the appropriate MSC. The MSC does not interface to the PSTN but makes sure that calls reach the correct subscriber within their coverage area.

2.5   The answer depends from which country you dial. In short, you would probably get a number unavailable tone. Outside of the United States, the switch would analyze the number and, since the number did not start with 00, would assume that it was a national call. It would then try to find the area code corresponding to 0331 (or 03311 if five-digit area codes are in use) and would probably fail to find a valid area. If, by chance, there was a local code 0331, it would try to connect you with the subscriber with a phone number 123456 in that local area. Within the United States such a number would be immediately detected as invalid.

3.1   12 minutes is 20% of an hour ($12/60 = 0.2$), hence 0.2E.

3.2   1 MHz or 10^6.

3.3   The key factors are radio spectrum assignment, traffic per user, and number of cells. Only the number of cells can be fully controlled by the operator, although traffic levels can be modified slightly through different tariffs.

3.4   Trunking gain is the phenomenon that the more channels that can be placed in a pool, the greater the efficiency with which each channel can be used for the same blocking probability. The implications are that operators should try to get as many channels in each cell by reducing the cluster size, getting as much spectrum as possible from the government, and not reserving spectrum for any particular users.

3.5 A small cluster size is best because it increases the number of radio frequencies available in each cell. The cluster size is limited by the distance traveled before the signal strength drops to a level where it will not cause interference.

4.1 The mobile would report its location to the network much less frequently; however, paging messages would need to be sent in many more cells.

4.2 Since the previous single cell was in one location area and the users will not be moving through the location area with any change in speed, then in principle all the new cells should be placed in the same location area as the old cell.

4.3 So the network can check that the mobile has not moved out of coverage, lost battery power, or malfunctioned and allow databases in the network to be rebuilt in the case of failure.

4.4 Because it first sends a detach message to inform the network that it is no longer turned on.

4.5 In the HLR (which may not be in the same network as the mobile if the mobile has roamed).

5.1 In all the cells in the location area where the mobile is known to be.

5.2 512, twice as many as with eight bits.

5.3 1) Paging message sent to the mobile. 2) Mobile responds with a random access message. 3) Network responds with a reserved channel. 4) Mobile sends its number to the network. 5) Network confirms the number back to the mobile. 6) Call is established.

5.4 Because as the traffic increases so does the number of collisions, resulting in repetition of random access attempts and a sudden increase in traffic. All that can be done is to progressively bar access classes until the traffic returns to manageable levels.

5.5   In case another mobile sent a random access attempt with the same random number at the same time. By sending one number back it uniquely identifies the mobile that has actually been allocated the channel.

6.1   In the network, although the information required to make the decision comes from the mobile.

6.2   Because the network is able to understand the load situation in all the cells and can make a better informed handover decision than the mobile would be able to.

7.1   01011001.

7.2   By putting the random number and the secret number into a special algorithm.

7.3   When the called party is in a different network from their home network and they are called by someone else in the same country as they currently reside in. If they are called by someone in a different country, then inefficient routing will also result, with the call completing two sides of a triangle when it could more efficiently be routed along the third side.

7.4   They do not. The secret number resides within the SIM card and under no circumstances can it be read directly from the card. The user has no need to know their own secret number.

7.5   From the first few digits of the mobile's telephone number, which contain the country code and network code of the home network.

8.1 A signal that is 20 dB stronger is 100 times stronger, one that is 3dB stronger is 2 times stronger. One that is 23dB stronger is $100 \times 2 = 200$ times stronger.

8.2 Fast fading is caused by the addition of more than one signal path, with the paths being of varying distance and hence varying phase, resulting in constructive and destructive additions as the mobile moves along. Fast fades are spaced at the same distance apart as the wavelength, hence 15 cm in this case.

8.3 In order to increase the range, reduce the fast fading and ISI and provide a small capacity increase.

8.4 Both are caused by reflections arriving later than the main path. In the case of ISI, the reflection arrives so much later that the main path is carrying the next transmitted bit. It stops becoming fast fading and starts becoming ISI when the extra delay on the longest significant reflected path is equal or greater to the time taken to send one bit.

8.5 Because the capacity of a cell is fixed regardless of its size, hence the more (small) cells the greater the overall system capacity.

9.1 The advantage is that they use less radio spectrum and hence more subscribers can be accommodated. The disadvantage is that they typically result in a lower voice quality than higher rate coders, although as coding techniques advance this is not always the case.

9.2 Interleavers are required because the radio channel is prone to errors occurring in bursts whereas error correction systems work best when errors occur evenly spread. The interleaver changes the order of transmission of bits so that if a burst of errors occurs, when the bits are reassembled in the correct order, this burst is spread across the received data.

9.3   First frequency hopping prevents a mobile from being stationary in a fade since fades are in different places for different frequencies. Second, frequency hopping increases capacity by reducing cluster size by distributing interference between mobiles more evenly and allowing frequencies to be reused closer together.

9.4   The transmitted sequence is spread by the spreading code, which results in the transmitted sequence having G times more bits than the data sequence, where G is the number of bits in the spreading sequence.

9.5   Orthogonal codes are such that when multiplied by each other the result is always zero, regardless of whether they are carrying a 0 or a 1 of user data. They are desirable because they minimize the interference between the users, resulting in the maximum possible capacity.

10.1   Capacity.

10.2   Around a 30% greater capacity in typical deployments.

10.3   These are typically differences in range, the effects of sectorization, the use of microcells, the operation in unlicensed bands and potentially in risk and cost. These are the key areas that need to be investigated when comparing TDMA and CDMA.

11.1   For two reasons: first, to provide a signalling burst for a second mobile when the channel is used for two half-rate mobiles, and second, to give the mobile time to listen to the control channel in adjacent cells.

11.2   Frequency correction burst to allow the mobiles to find the control channel. Synchronization burst to allow the mobiles to understand where they are in the TDMA burst structure. Broadcast information providing information specific to the cell. Paging information telling the mobile that there is an incoming call for it.

11.3 To minimize the amount that the channel has changed between the transmission of the sounding sequence and the transmission of the data.

11.4 Typically 12 for GSM and 1 for CDMA. They differ because CDMA can tolerate higher levels of interference from surrounding cells than can GSM.

11.5 A process whereby the mobile is in contact with two cells simultaneously while it moves between them.

12.1 Because providing coverage levels equivalent to cellular has proved too expensive and users are not prepared to tolerate lower coverage levels. Only in those cities where the numerous small cells can be justified on the grounds of requiring an exceptionally high capacity has telepoint succeeded.

12.2 It allows spatial diversity or antenna diversity to be implemented without the need for a second antenna on the mobile by taking advantage of the uplink and downlink having the same fast fading properties.

12.3 Mainly because the use of directional antennas at the subscriber unit strongly reduces the signal strength of reflections from paths off to one side or behind the receiver, thus reducing the ISI to a level where the system can operate satisfactorily. The directional antenna also increases the range by focusing the power in a narrow beam.

12.4 Very few. The differences mainly lie in the fact that PHS provides 4 channels on a TDMA bearer whereas DECT provides 12. PHS also has a slightly more advanced modulation scheme that allows it to provide a greater spectrum efficiency.

12.5 Because cordless systems have been deliberately designed to be simple and as a result do not have equalizers. The implication of this is that cordless systems can only work over a short range unless deployed in a WLL configuration.

13.1 Just like cellular, the basic components are switching, base stations, and receiver terminals, which in this case are not mobile.

13.2 The key differences revolve around the fact that WLL subscribers do not move. WLL networks do not require location databases, for example. However, they do need to provide better voice quality at a lower cost. Finally, WLL networks can connect directly into a PSTN switch rather than having their own special switch (although most have their own switch, which is similar to that provided by the PSTN).

13.3 ADSL over copper cable, cable modems, and video distribution systems.

13.4 Because the cost of installing a radio unit on a home is much lower than the cost of digging up the road, particularly where only a few subscribers want the service so that a lot of road needs to be dug up between each subscriber.

13.5 They provide very high data rates to the subscriber but only low data rates back to the network. They are at higher frequencies, so they have a lower range. Other than this they are very similar.

14.1 Basically, that the system designed for WLL has base stations that connect directly to the PSTN switch rather than to an MSC. Also, HLRs and VLRs are no longer required.

14.2 Both. It is a framing format for data transmission that can be used over any medium.

14.3 Because such provision is incredibly inefficient of radio spectrum resource. It would require around 20 times as much radio spectrum. The only advantage would be that subscribers would never be blocked whereas when a permanent line is not provided there is a small probability that they might be blocked.

16.1 That the satellite stays in the same position relative to the Earth and so directional antennas do not need to be able to track the satellite.

16.2 1) Because the resulting cells are too big to provide sufficient capacity. 2) Because the delay is too great. 3) Because the transmit power required would rapidly drain a mobile's battery.

16.3 Between 500 and 1,000 km.

16.4 It would search for a GSM signal and, if it found one, would use the GSM service. If it was unable to find a GSM signal it would switch automatically to Iridium, occasionally searching for a GSM network and switching to it whenever it was detected.

Glossary

Access technology   A means of connecting a home into the PSTN or of providing it with access to the PSTN.

ADPCM   Adaptive Differential Pulse Code Modulation. A type of speech coding using 32 Kbits/s to transmit voice. It is widely used in WLL networks and cordless systems.

AM   Amplitude Modulation. A type of modulation where the amplitude of the transmitted wave is varied in accordance with the user signal.

Amplitude Modulation   See AM.

Antenna diversity   See diversity.

ARQ   Automatic Repeat Request. A type of error correction system whereby when errors are detected the receiver requests that the data be resent.

ASCI   Advanced Speech Call Items. The addition of group and broadcast calls to the GSM standard.

Associated control channel   A control channel associated with a particular voice channel.

AuC   Authentication Center. The part of the GSM network that holds secret information about the mobile and is able to confirm that the mobile user is who they claim to be.

ADSL   Asynchronous Digital Subscriber Line. A technique for transmitting very high data rates down the existing copper cables to the home.

Attach   The process of the mobile informing the network that it is on so that calls can be forwarded to it.

Authentication center   See AuC.

Bandwidth   The amount of radio spectrum required to send a radio signal. See the more detailed explanation in Section 8.2.

Base station   The transmitter/receiver that sends radio signals to the mobiles.

Base station controller   See BSC.

Base station subsystem   See BSS.

Bearers   The copper wire or radio channel that is used to carry the signal from the user into the network, and vice versa.

Bel   See dB.

BER   Bit Error Rate. The portion of the transmitted bits that are received in error as a result of interference and fading on the radio channel.

Bit error rate   See BER.

Bits   Single binary digits. See the explanation in Section 5.3.

Block coding   A type of error correction system where the original data and the result of a matrix multiplication are sent to the receiver.

Broadband   The sending of data requiring a wide bandwidth. Although not well defined, it typically refers to data requiring a wider bandwidth than voice, for example, video telephony and high-speed file transfer.

BSC   Base Station Controller. The part of a GSM system that controls a number of base stations and acts as a node for connecting base stations to the switch.

BSS   Base Station Subsystem. The combination of base stations and BSCs that together provide the radio functionality in a cellular system.

Burst   A continuous transmission by a mobile using TDMA lasting the period of a slot.

Bytes   A collection of eight bits. See bits.

CAMEL   Customized Applications for Mobile Enhanced Logic. A terrible acronym, this means the addition of intelligent network capabilities to the GSM system.

Carrier   A single radio frequency band. In a TDMA system this contains a number of channels. For example, in the GSM system, each TDMA carrier contains eight channels.

CATU   Central Access and Transcoding Unit. The part of the Lucent AirLoop WLL system that connects to the switch.

CDMA   Code Division Multiple Access. The process of letting all the users transmit on the same frequency at the same time but distinguishing between them by the code used.

Cellular radio   The use of a number of separate radio cells to provide mobile radio telephony over a large area to a large number of subscribers but using a relatively small number of radio frequencies and reusing them in different cells in such a way that interference is minimized.

Clipping   The loss of the first syllable of a word when discontinuous transmission is used.

Cluster   A set of cells using different radio frequencies so that they do not interfere with each other. The set of cells can be repeated indefinitely across a large area. See Section 3.3.

Channel   The radio resources used by a single user. In the case of FDMA, this is equal to a carrier. In the case of TDMA, it is equal to part of a carrier.

Circuit switching   The establishment of a dedicated channel for the duration of a call. No other mobiles are able to use that channel until the call is finished. See also packet switching.

Code division multiple access   See CDMA.

Control channel   A channel used for the purposes of controlling the mobile and not for sending any of the user's data.

Convergence, fixed/mobile   The concept that fixed and mobile phones will be replaced by a single phone with the functionality of both.

Convolutional coding   A type of error correction coding where the data to be transmitted is substantially modified in the process of adding redundancy.

Cordless   A type of mobile phone designed for home and office use with a limited range and typically a simple design.

CTRU   Central Transceiver Unit. The base station in the Lucent AirLoop WLL system.

dB   Decibel. A tenth of a Bel. A way of writing the relative size of two numbers without needing to write numerous zeros before the numbers. See the more detailed explanation in Section 8.1.

DCA   Dynamic channel allocation When the transmitter does not have fixed frequencies assigned to it but selects the frequency with the lowest interference at the time when a user makes a call. See Section 10.2.

Decibel   See dB.

DECT   Digital Enhanced Cordless Telephone. A European standard for a cordless phone.

Detach   The process of the mobile informing the network that it has been turned off so that calls are no longer sent to it.

Discontinuous transmission The stopping of the transmission of a radio signal when the user is not actually talking, for example, when they are listening to the other person. For a more detailed explanation see Section 10.3.

Diversity The use of two antennas to receive the same signal so that when one antenna is in a fade, typically the other antenna is not and the resulting received signal is enhanced. See the more detailed explanation in Section 12.2.

Downlink Transmissions from the base station to the mobile.

DS-CDMA Direct Sequence CDMA. The form of CDMA normally considered as normal CDMA (as opposed to FH-CDMA). See CDMA.

Dual-band Phones that operate with only one standard but are capable of working in two different frequency bands.

Dual-mode Phones capable of operating on more than one standard.

Dynamic channel allocation See DCA.

EIR Equipment Identity Register. The part of the GSM system that retains information about whether the mobile has been stolen.

Encryption The process of scrambling a radio signal so that eavesdroppers cannot understand it. It relies on modifying a digital signal by a mask that only the transmitter and the receiver know.

Equipment identity register See EIR.

Equalization The process of removing ISI.

Erlangs A measure of radio traffic. 1 Erlang is equivalent to 1 telephone line being permanently used. See Section 3.4.

Error correction The process of removing errors in the received signal that have been caused by loss of radio signal due to fading or other propagation problems. The receiver uses redundancy added into the transmitted signal to correct the errors.

Fast fading   The rapid change in signal, with the signal falling almost to zero and then increasing back again every wavelength of distance travelled, typically some 30 cm in cellular systems. It is caused by reflected waves having different phases from the main signal path.

FDMA   Frequency Division Multiple Access. The division of the available spectrum into small frequency bands that are given to each mobile.

FH-CDMA   Frequency Hopped CDMA. A type of CDMA where instead of spreading the signal by the spreading code the user moves rapidly from frequency to frequency.

Frequency division multiple access   See FDMA.

Frequency hopping   The process of rapidly changing from frequency to frequency to reduce the effect of fades and to spread interference more evenly to all mobiles.

GAP   Generic Access Protocol. A protocol used by DECT to ensure compatibility between equipment from different manufacturers.

Gateway mobile switching center   See GMSC.

GHz   Thousands of millions of hertz. See Hz.

GMSC   Gateway mobile switching center. The part of the GSM network responsible for providing a point of connection between the GSM network and the PSTN. Incoming calls are sent to the GMSC, which then routes them to the mobile.

GPRS   General Packet Radio Service. The addition of packet data capabilities to the GSM standard.

GSM   The Global System for Mobile communications. A particular type of cellular system now widely implemented throughout the world.

Half-rate speech coder   A speech coder for GSM that can send speech using only half as much bandwidth as the current speech coder. The speech coder has now been developed but is not widely used because operators have concerns over its quality.

Handoff   The process of transferring a mobile from one cell to another while it is in the middle of a call.

Hierarchical cell structure The result when microcells are deployed in an area covered by larger cells. Most areas will have coverage from both the large cell—called the macrocell, oversailing cell, or umbrella cell—and the microcell.

HSCSD High-Speed Circuit-Switched Data. The provision of data rates of up to 64 Kbits/s to the GSM standard.

Hz Hertz. The number of times that a sine wave repeats in a second. See the more detailed explanation in Section 3.2.

HLR Home Location Register. The part of the GSM system responsible for remembering the location of a mobile. A mobile's HLR is always situated in the network where it has its subscription.

Home location register See HLR.

IMEI International Mobile Equipment Identity. The serial number of the mobile that is used to check for stolen mobiles.

IMSI International Mobile Subscriber Identity. The number used within GSM to contact mobiles. This is the personal telephone number that the mobile knows.

IMT2000 The international vision of third generation mobile services.

IN Intelligent Network. A concept whereby it is easy to add new service to the network using software. See the more detailed explanation in Section 17.2.

Information Society The concept that the developed world is increasingly becoming a place where users exchange vast amounts of information.

Intelligent Networks See IN.

Interleaving A technique for reordering the data to be transmitted so that bursts of errors on the channel result in errors that are more evenly distributed after deinterleaving and, hence, simpler to correct.

Internet telephony The sending of telephone calls over the Internet, rather than the PSTN, to take advantage of the lower call costs on the Internet.

Intersymbol Interference   See ISI.

ISDN   Integrated Services Digital Network. A framing format allowing flexible transmission of digital information across a range of bearers. See the more detailed explanation in Section 14.2.

ISI   Intersymbol Interference. The resulting effect when one reflected path has a much longer delay than the direct path, resulting in the signal arriving on the reflected path while the information relating to the next bit is being received on the direct path.

Kbits/s   Thousands of bits per second, or bits/s × 1000.

kHz   Thousands of hertz, or Hz × 1000.

km   Kilometers. Each kilometer is 1,000m.

LEO   Low Earth Orbit. The orbit used by personal satellite communications services to allow sufficient subscriber capacity and reduce delay and battery consumption.

Line-of-sight   See LOS.

Local loop   The connection from the switch to the subscriber's home. It is called "loop" because each connection requires two wires that look like a loop when drawn on a diagram and "local" to distinguish the connections to homes from connections made between different switches.

LOS   Line-of-sight. A direct radio path between the transmitter and the receiver, without any reflections or diffraction. This occurs when the receiver can be visually seen from the transmitter.

Low earth orbit   See LEO.

LMDS   Local Multipoint Distribution System. A new digital video distribution system proposed for the United States.

Location area   A collection of cells in the same area. When a mobile moves into a new location area it must report its location. Hence its location is only known to the accuracy of which location area it is in.

Logarithm   See dB and Section 8.1.

Macrocell   See hierarchical cells.

Mbits/s Millions of bits per second. See bits.

MHz Millions of hertz. See Hz.

Microcells Small cells that result when the base station antenna is placed below the height of the surrounding rooftops. Such cells are typically less than 500m in length and follow the local street pattern.

Microwave link A radio link between two fixed points using directional antennas in order to provide communications. It is often used to link a base station back to a switch.

MMDS Microwave Multipoint Distribution System. A type of video distribution system used in the United States.

Mobile switching center See MSC.

Modulation The process of changing a high-frequency radio wave in accordance with the speech or data information to be transmitted. In the simplest form, the amplitude of the high frequency or carrier wave is varied in accordance with the amplitude of the speech signal. See AM, FM, PM, and Section 9.5.

MSC Mobile Switching Center. The switch in a mobile network that connects calls from the GMSC to the cell where the mobile is located. The MSC also manages the handover function and analyzes call origination messages.

MSISDN Mobile Station Integrated Services Digital Number. The number dialed when someone wishes to call a GSM mobile.

MVDS Microwave Video Distribution System. A type of video distribution system adopted in Europe.

Narrowband The transmission of signals only requiring a narrow bandwidth. Although not well defined, this typically refers to voice traffic and voiceband data traffic.

OMC Operations and Maintenance Center. The part of the network where faults are reported and statistics gathered on network performance.

Operations and maintenance center See OMC.

Orthogonal codes   Codes that, when multiplied by each other, result in zero, minimizing interference between mobiles using CDMA transmission.

Oversailing cell   See hierarchical cells.

NIU   Network Interface Unit. The subscriber equipment in the Lucent AirLoop WLL system.

PABX   Private Automatic Branch Exchange. The telephone exchange typically found in most offices that performs the switchboard function.

Packet switching   The segmentation of the data or speech generated by the user into packets of bits, each typically only a few milliseconds long. These packets are then transmitted at regular intervals on whatever channel is available when they are transmitted. See also circuit switching.

Paging   The sending of a message to a mobile to inform it that there is an incoming call for it.

Pathloss   The reduction in signal strength from that transmitted to that received, which results from sending the radio signal from the transmitter to the receiver.

PCM   Pulse Code Modulation. A type of speech coding that requires 64 Kbits/s of data to transmit the voice information. (An 8-kHz sampling rate so that a typical 3-kHz audio signal can be reproduced and eight bits to indicate the signal amplitude.) It is widely used in fixed networks.

Phase modulation   See PM.

PHS   Personal Handiphone System. The Japanese cordless telephone standard.

PM   Phase Modulation. A type of modulation whereby the phase of the transmitted signal is varied in accordance with the user data.

PMR   Private Mobile Radio. The use of radio for professional purposes where the radio transmitters are owned by the users. Typical examples are police and railways.

PN Pseudo Noise. A type of code used in CDMA that is near-orthogonal and has the advantage of having a huge number of different codes in the codeset.

Power control The variation of the mobile's power depending on the distance from the transmitter so that signals are received with an adequate but not excessive power. See Section 10.3.

PSTN The Public-Switched Telephone Network. The worldwide network, consisting of copper cables and switches in most countries in the world, which allows you to call anywhere in the world. See the more detailed explanation in Section 2.3.

Pseudo noise See PN.

Public Switched Telephone Network See PSTN.

Pulse code modulation See PCM.

Radio frequency See RF.

Random access channel A radio channel reserved for mobiles to use when they wish to make contact with the network, typically either to originate a call or to respond to a paging message.

Rayleigh fading See fast fading.

Radio spectrum See spectrum.

Repeater units The use of base stations to receive a signal and to re-broadcast the same signal, typically on a different frequency. By this means the range of a cell can be increased although spectrum efficiency is reduced.

RF Radio Frequency. The parts of the mobile and base station that deal with the high-frequency signal used for transmission that results after modulation. More generally, the higher frequencies (megahertz and gigahertz) used in radio transmission as opposed to the lower frequencies (kilohertz) found in speech signals.

Roaming Taking your cellular phone to a different country from the one where you have a subscription (i.e., the one where the operator to whom you pay your bill is located).

RPE-LTP Regular Pulse Excited—Long-Term Prediction. The type of speech coder used in GSM. It uses 13 Kbits/s to represent voice.

Sectorization The division of a circular cell into a number of pie-shaped cells using directional antennas at the base station.

Signal-to-Noise Ratio See SNR.

Signalling information Information sent to or from the mobile associated with keeping the call functioning correctly (as opposed to the voice signal sent on the audio channel).

SIM Subscriber Identity Module. A card within the GSM mobile that carries all the information specific to the user of the phone.

SIM card roaming The taking of the SIM card to a different country where normal GSM terminals cannot be used, and the placement of the SIM card in a modified GSM terminal that will work in the new country.

Slow associated control channel A GSM associated control channel used for the transmission of information to mobiles relating to signal strength in surrounding cells and handover.

Slow fading The loss in radio signal as a mobile passes behind buildings and other large obstacles.

SMS Short Message Service. The capability to send a short text message of up to 160-characters long to a GSM phone.

SNR Signal-to-Noise Ratio. A measure of the difference between the wanted radio signal and the background noise in the channel.

Sounding sequence A short burst of bits that is sent from the transmitter to the receiver in order to allow it to estimate the echoes in the radio channel and hence set up the equalizer appropriately to cancel the ISI that would otherwise occur.

Spatial diversity See diversity.

Spectrum Spectrum basically means a sequence of frequencies. Each radio signal needs a certain amount of these frequencies ("bandwidth") to permit the transmission of information from the transmitter to the receiver. Spectrum is a finite quality like land and hence is very valuable.

S-PCN   Satellite Personal Communications Networks.

TDD   Time Division Duplex. The transmission of the downlink and the uplink on the same frequency, but at different times, in a TDMA format.

TDMA   Time Division Multiple Access. The process of dividing up the available spectrum in time and giving all the spectrum to each user for a short period of time (timeslot).

Telepoint   The use of cordless phones in a cellularlike application to provide telephone coverage in extremely densely populated cities. Most telepoint services have failed.

Time division duplex   See TDD.

Time division multiple access   See TDMA.

Timeslot   See TDMA.

Timing advance   The transmission of a burst in a TDMA system before the slot is ready at the base station to allow for the delay in the radio signal reaching the base station. See Section 11.3.1 for a more detailed explanation.

Traffic channel   See channel.

Transceiver   A unit that combines the function of transmitter and receiver as in a cellular base station and, indeed, a mobile.

Trombone   The inefficient route taken by a call in a GSM network when the subscriber has roamed to another country and someone calls him from that other country.

Trunking efficiency   The phenomenon that, as more radio channels and users are clubbed together, the efficiency of use of each radio channel increases.

TRX   Transmitter/Receiver. A common abbreviation used by cellular engineers. See transceiver.

Uplink   Transmission from the mobile to the base station.

Umbrella cell   See hierarchical cells.

UMTS Universal Mobile Telecommunications Service. One name for the third-generation mobile radio system. Typically this refers to the European vision of third-generation mobile.

V5 An interface standard to connect WLL systems to the PSTN switch. There are currently two variants, V5.1 and V5.2.

Voice activity detector A device that decides whether the user is actually speaking used to implement discontinuous transmission.

Video distribution systems Radio systems with a similar architecture to WLL but operating at much higher radio frequencies and capable of transmitting much more data. They are used to provide 50 or so television channel to nearby homes.

Viterbi decoder A complex decoding system that can be used for optimally decoding convolutional codes and also for removing ISI from a signal.

Visitors location register See VLR.

VLR Visitors Location Register. The part of the GSM system responsible for keeping track of a mobile's position to the nearest location area.

Walsh codes A family of orthogonal codes often preferred for CDMA transmission.

WLL Wireless Local Loop. The use of radio to replace copper wiring as a means of connecting the home to the PSTN.

About the Author

WILLIAM WEBB graduated in electronic engineering with a first class honors degree and all top year prizes in 1989. In 1992 he earned his Ph.D. in mobile radio and in 1997 he was awarded an MBA.

From 1989 to 1993 William worked for Multiple Access Communications Ltd. as the Technical Director in the field of hardware design, modulation techniques, computer simulation, and propagation modeling. In 1993 he moved to Smith System Engineering Ltd. where he was involved in a wide range of tasks associated with mobile radio and spectrum management. In 1997 he moved to Netcom Consultants where he lead the wireless access division and in 1998 to Motorola as a director involved in cellular and WLL.

William has published over forty papers, holds four patents, and was awarded the IERE Premium in 1994. He is the author of *Modern Quadrature Amplitude Modulation*, published by John Wiley, and *Wireless Local Loop*, published by Artech House.

Index

The Artech House Mobile Communications Series

John Walker, Series Editor